THE VCR AGE

SOME OTHER VOLUMES IN THE SAGE FOCUS EDITIONS

8. **Controversy (Second Edition)**
   Dorothy Nelkin
21. **The Black Woman**
    La Frances Rodgers-Rose
31. **Black Men**
    Lawrence E. Gary
32. **Major Criminal Justice Systems (Second Edition)**
    George F. Cole, Stanislaw J. Frankowski, and Marc G. Gertz
41. **Black Families (Second Edition)**
    Harriette Pipes McAdoo
51. **Social Control**
    Jack P. Gibbs
57. **Social Structure and Network Analysis**
    Peter V. Marsden and Nan Lin
64. **Family Relationships in Later Life**
    Timothy H. Brubaker
65. **Communication and Organizations**
    Linda L. Putnam and Michael E. Pacanowsky
66. **Competence in Communication**
    Robert N. Bostrom
67. **Avoiding Communication**
    John A. Daly and James C. McCroskey
68. **Ethnography in Educational Evaluation**
    David M. Fetterman
71. **The Language of Risk**
    Dorothy Nelkin
72. **Black Children**
    Harriette Pipes McAdoo and John Lewis McAdoo
74. **Grandparenthood**
    Vern L. Bengtson and Joan F. Robertson
75. **Organizational Theory and Inquiry**
    Yvonna S. Lincoln
76. **Men in Families**
    Robert A. Lewis and Robert E. Salt
77. **Communication and Group Decision-Making**
    Randy Y. Hirokawa and Marshall Scott Poole
78. **The Organization of Mental Health Services**
    W. Richard Scott and Bruce L. Black
79. **Community Power**
    Robert J. Waste
80. **Intimate Relationships**
    Daniel Perlman and Steve Duck
81. **Children's Ethnic Socialization**
    Jean S. Phinney and Mary Jane Rotheram
82. **Power Elites and Organizations**
    G. William Domhoff and Thomas R. Dye
83. **Responsible Journalism**
    Deni Elliott
84. **Ethnic Conflict**
    Jerry Boucher, Dan Landis, and Karen Arnold Clark
85. **Aging, Health, and Family**
    Timothy H. Brubaker
86. **Critical Issues in Aging Policy**
    Edgar F. Borgatta and Rhonda J.V. Montgomery
87. **The Homeless in Contemporary Society**
    Richard D. Bingham, Roy E. Green, and Sammis B. White
88. **Changing Men**
    Michael S. Kimmel
89. **Popular Music and Communication**
    James Lull
90. **Life Events and Psychological Functioning**
    Lawrence H. Cohen
91. **The Social Psychology of Time**
    Joseph E. McGrath
92. **Measurement of Intergenerational Relations**
    David J. Mangen, Vern L. Bengtson, and Pierre H. Landry, Jr.
93. **Feminist Perspectives on Wife Abuse**
    Kersti Yllo and Michele Bograd
94. **Common Problems/Proper Solutions**
    J. Scott Long
95. **Falling from the Faith**
    David G. Bromley
96. **Biosocial Perspectives on the Family**
    Erik E. Filsinger
97. **Measuring the Information Society**
    Frederick Williams
98. **Behavior Therapy and Religion**
    William R. Miller and John E. Martin
99. **Daily Life in Later Life**
    Karen Altergott
100. **Lasting Effects of Child Sexual Abuse**
     Gail Elizabeth Wyatt and Gloria Johnson Powell
101. **Violent Crime, Violent Criminals**
     Neil Alan Weiner and Marvin E. Wolfgang
102. **Pathways to Criminal Violence**
     Neil Alan Weiner and Marvin E. Wolfgang
103. **Older Adult Friendship**
     Rebecca G. Adams and Rosemary Blieszner
104. **Aging and Health**
     Kyriakos S. Markides
105. **The VCR Age**
     Mark R. Levy
106. **Women in Mass Communication**
     Pamela J. Creedon
107. **Midlife Loss**
     Richard A. Kalish
108. **Cultivation Analysis**
     Nancy Signorielli and Michael Morgan
109. **Black Students**
     Gordon LaVern Berry and Joy Keiko Asamen

THE VCR AGE
Home Video and Mass Communication

Edited by
Mark R. Levy

SAGE PUBLICATIONS
The Publishers of Professional Social Science
Newbury Park   London   New Delhi

Copyright © 1989 by Sage Publications, Inc.

All rights reserved. No part of this book may be reproduced or utilized in any form or by any means, electronic or mechanical, including photocopying, recording, or by any information storage and retrieval system, without permission in writing from the publisher.

For information address:

SAGE Publications, Inc.
2111 West Hillcrest Drive
Newbury Park, California 91320

SAGE Publications Ltd.
28 Banner Street
London EC1Y 8QE
England

SAGE Publications India Pvt. Ltd.
M-32 Market
Greater Kailash I
New Delhi 110 048 India

Printed in the United States of America

Library of Congress Cataloging-in-Publication Data

Main entry under title:

The VCR age : home video and mass communication / Mark R. Levy, editor.
    p.   cm. — (Sage focus editions ; v. 105)
   Bibliography: p.
   ISBN 0-8039-3299-5. — ISBN 0-8039-3300-2 (pbk.)
   1. Video tape recorder industry. 2. Video tape industry.
3. Videocassettes. 4. Market surveys. 5. Mass media. I. Levy, Mark R.
HD9696.V532V37   1989
338.4'7621388332—dc19   89-5852
                                                                                   CIP

FIRST PRINTING 1989

Contents

Introduction

1. Why VCRs Aren't Pop-Up Toasters: Issues in Home Video Research   9
   MARK R. LEVY

PART I: The Growth of Home Video

2. The Diffusion of the VCR in the United States   21
   BRUCE C. KLOPFENSTEIN
3. Home Video: The Consumer Impact   40
   PAUL B. LINDSTROM
4. The Uses and Impact of Home Video in Great Britain   50
   BARRIE GUNTER and MALLORY WOBER

PART II: Using the VCR

5. Adolescents and the VCR Boom: Old, New, and Nonusers   73
   BRADLEY S. GREENBERG and CAROLYN LIN
6. Social and Psychological Antecedents of VCR Use   92
   ALAN M. RUBIN and REBECCA B. RUBIN
7. Subjective Differences in the Use and Evaluation of the VCR   112
   MILTON J. SHATZER and THOMAS R. LINDLOF

PART III: The VCR and the Individual

8. Big Eyes But Clumsy Fingers: Knowing About and Using Technological Features of Home VCRs   135
   AKIBA A. COHEN and LAURA COHEN
9. Measuring VCR "Ad-Voidance"   148
   BARRY S. SAPOLSKY and EDWARD FORREST
10. School Achievement, Self-Esteem and Adolescents' Video Use   168
    KEITH ROE

PART IV: VCRs, Groups, and Societies

11. Away from the Mainstream? VCRs and Ethnic Identity 193
    JULIA R. DOBROW
12. VCR Narrowcasting in the Kibbutz 209
    DOV SHINAR
13. The Worldwide Cultural and Economic Impact of Video 230
    CHRISTINE OGAN
14. The Videocassette Recorder in the USSR and Soviet-Bloc Countries 252
    DOUGLAS A. BOYD

About the Contributors 271

Introduction

1

Why VCRs Aren't Pop-Up Toasters: Issues in Home Video Research

MARK R. LEVY

Recently, two colleagues and I submitted an article about home video cassette recorders (VCRs) to a major communication journal. After an uncharacteristically long delay between submission and response, we received "The Word" from the editor: Revise and resubmit your article, taking into account the referee's comments, especially Reader #2.

Now like most scholars, I have received my share of helpful reviews—referee's comments that vastly improved the quality of the finished article. And, of course, I have received my share of reviews obviously scribbled by narrow-minded fools.

On first glance, I thought Reader #2 fell into this latter category, one more example of a dull-normal pedant who wouldn't recognize a good idea if it socked him/her in the snoot.

Here's what Reader #2 said:

> A path of VCR studies could be important. Yet you have the obligation to establish what type of "important role" the VCR plays and why it is significant to study. Is it only significant because of widespread diffusion (i.e., because it's there)? That could lead to a rash of studies of electric toasters....

Electric toasters? Give me a break. Why not electric coffee pots or electronic water-piks while you're at it? Just how nasty can we be, I wondered, when we have the anonymous review to hide behind.

Some time has now passed. In due course, the revised article was published, and my anger has cooled to the point that I could *even* see some wisdom in Reader #2's sarcastic comments. Why indeed is it "significant to study" home video? What *is* the place of the home videocassette recorder in mass communication?

In this chapter, I want to review some of what we know, and don't know, about home video; to introduce the baker's dozen of specially commissioned essays that make up the bulk of this book; and to suggest some future directions for VCR research. In the process, I hope to be able to answer Reader #2's challenge and to demonstrate that VCRs are something much more than pop-up toasters.

First, let me dispose of the "Mt. Everest" problem. Is it sufficient justification to study VCRs *simply* because, like Everest, they are there? Of course not. Scholars aren't mountain climbers, although they certainly are (or should be) explorers of a sort. Rather, I'd propose a simple *Gedankenexperiment*. What if thirty-five years ago, I had submitted a research paper to a journal, dealing with what was then the new home communication technology—television. And what if a referee then had asked, "Why investigate the role of *television* in the mass communication process, after all it's nothing more than another electrical appliance?"

A scientifically justified response then—and now—would be, "I don't know. I have a hunch that something significant is happening. But we won't know, unless we do the research." Not to belabor the point; but, if, as is the case, no other communication technology since television has achieved levels of *home* penetration comparable to the VCR; if dramatically more than half of all U.S., Japanese, and British households now have VCRs; and if video ownership in the Arabian Gulf states and much of the Third World is well above that mark, it strikes me as either the height of Himalayan myopia or the worst case of penetration envy ever to ignore home video.

THE GROWTH OF HOME VIDEO

What then do we know about the VCR Age? For some time now, researchers have wondered, for example, whether home video heralds an evolutionary or revolutionary development in the way people use communication media. While a few investigators have suggested that VCRs can be plausibly characterized as some sort of radical discontinuity in the mass communication process (see, for example, Baboulin, Gaudin & Mallein, 1983; Gubern, 1985; Cubbitt, 1986), most students of this new technology (e.g., McQuail, 1986; Rubin & Bantz, 1987; Levy & Gunter, 1988) have suggested that the VCR is largely an extension of and complement to older technologies.

Overall, I would judge the evolution/revolution debate to be sterile or at best nonproductive. It has, with few exceptions, failed to provide very many sophisticated insights about how to measure media revolutions, let alone how to apply those standards to gauging the role of VCRs.

Lacking good theory, however, we do have some important data. It seems clear from my reading of the research literature that home video has failed in at least two ways to make a revolution:

(1) From the perspective of the mass audience, home video has not achieved any special priority in terms of its perceived utility to gratify various needs (see, for example, Cohen, Levy & Golden, 1987) and has thus not established an identity separate from television and movies. Both conditions—a substantial ability to gratify media-related needs and a separate public identity—are arguably part of any evidence that would support a "revolutionary" conclusion about this new technology and its users.

(2) From the perspective of the mass media marketplace (see, for example, Kelly, 1988) home video has, by and large, not lived up to its potential to provide an increased range and diversity of program content. So-called special interest video (e.g., high culture, video art, political and social documentaries) has been unable to create a large enough audience to be economically profitable; and other, nonmarket mechanisms for financing home video narrowcasting have not been forthcoming.

Revolutionary technology or not, the VCR still may have important consequences for mass (and not-so-mass) communication. The home video cassette recorder represents a target of opportunity for mass communication researchers, for at the very least its newness provides a fresh research site in which to examine some of the enduring themes and controversies of media theory.

Considerable research, for example, has already focused on such questions as the frequency, type, and origins of programs recorded, replayed, rented, or bought (for early examples, see Levy, 1983; Darkow, 1985; Tydeman & Kelm, 1986). There is a general consensus in this literature that most VCR use in the U.S. and western Europe is a behavioral complement to existing patterns of television exposure and that the net effect of video ownership is only a slight, if any, diminution of live broadcast viewing. Moreover, VCRs in the U.S. and

England are reported to be used in roughly equal amounts for time-shifting or playback of rented/purchased prerecorded cassettes; while, by contrast, patterns of video use in other national settings—the Middle East or Third World, for example—reportedly involve much higher levels of prerecorded tape replay and virtually no time-shifting (see, for example, Boyd, 1987).

It should also be noted that to date little evidence has been accumulated on: (1) long-term patterns of video use (Does time-shifting, for example, decline as the "novelty" wears off?); (2) the so-called "library-building" function (Compared to music, for example, how many times will a VCR owner watch his or her favorite tape?); and (3) connoisseurship (the use of video by individuals who use the VCR to cultivate a highly sophisticated and critical understanding of film and television genres or *auteurs*.

Part I of this book builds on, synthesizes, and extends this early literature about how the public uses home video. In "The Diffusion of the VCR in the United States," Klopfenstein outlines the history of home videocassette recorder technology; offers year-by-year data, detailing the phenomenal and somewhat unpredicted adoption of the home videocassette recorder; and suggests likely trends for future increases in VCR ownership. The VCR, Klopfenstein concludes, "has become an integral part of the U.S. television household."

Just how "integral" is a matter of some controversy among communication scholars and industry researchers. In "Home Video: The Consumer Impact," Paul Lindstrom of Nielsen Media Research examines patterns of VCR recording, playback, and prerecorded video rental. Lindstrom contends that VCR use represents only a small portion of the time most households spend with television, and, writes Lindstrom, it is largely a myth that home video "will destroy pay television . . . bankrupt the networks. . . and create a generation of 'couch potatoes.'"

Picking up on a methodological theme from the Lindstrom chapter, Gunter and Wober address the question of how different measurement techniques (diaries, self-reports, people-meters) affect what we know about VCR use. In their chapter, "The Uses and Impact of Home Video in Great Britain," Gunter and Wober also examine gender differences in VCR use, whether home video promotes "family togetherness," and to what extent having a VCR makes watching TV more meaningful and enjoyable.

USING THE VCR

Chapters in Part II of this book focus on specific patterns of home video use and the social and psychological factors that condition and influence those behaviors. At the core of these chapters is a concern with "common sense" theory (McQuail, 1987) about home video. In brief, common sense theory is what the public thinks the VCR is good for. Early on (see, for example, Buckwalter, 1978; Consumers Reports, 1980) people apparently decided that the VCR held out three major communication possibilities: first, as a device to record television programs for later, more convenient replay (time-shifting); second, as a means for building a home library of previously broadcast programs and prerecorded tapes; and third, as a way to play rented or purchased tapes of movies, rock videos, and the like.

While for some of its users the VCR is apparently transparent, an unseen device of little significance in itself, other VCR users actively seek and obtain gratifications from this technology (e.g., to feel in greater control of the communication process or to feel up-to-date and modern) which go beyond those gratifications associated with content (Williams, et al., 1985; Schoenbach & Hackforth, 1987; Roe, 1987).

Since VCR-oriented gratifications are based on learned expectations about that technology, it is also relevant for researchers to ask where the public's perceptions of video originate in the first place (advertising, the media, interpersonal channels or whatever), and what are the general processes of video adoption? Does it make any difference to our understanding of the diffusion process, for example, that the innovation (video) itself is a means of communication? And what can VCR research tell us about the socialization of children and the resocialization of adults to the media and communication technology?

For example, in Chapter 5, "Adolescents and the VCR Boom: Old, New and Nonusers," Greenberg and Lin investigate the "increasingly rich media environment" of today's teenagers. So commonplace has the VCR become that there are, Greenberg and Lin observe, strikingly few differences between old, new, and non-VCR households.

Building on the uses and gratifications perspective, Rubin and Rubin ("Social and Psychological Antecedents of VCR Use") identify five principal motives for using home video, including *interpersonal* needs for social affiliation and affection. Social demography, psychological predispositions, and life position all influence what Rubin and Rubin call "the purposeful and active uses of VCRs."

In "Subjective Differences in the Use and Evaluation of the VCR," Shatzer and Lindlof tackle similar issues to those dealt with by Rubin and Rubin. In their chapter, Shatzer and Lindlof identify three types of VCR users (the polychronic, the price conscious, and the tape-renters) and they show that life-styles, less than demography, influence whether VCR users seek to "maximize" or "satisfice"in their video usage.

THE VCR AND THE INDIVIDUAL

The next three chapters, which are grouped in Part III, take as their unit of analysis the individual, and ask about specific VCR behaviors and their consequences. What people know about the actual hardware of home video and whether they use all the available bells and whistles is discussed in Chapter 8, "Big Eyes but Clumsy Fingers: Knowing About and Using Technological Features of Home VCRs." The authors, Cohen and Cohen, conclude that there is "less than perfect knowledge about the VCR and far from full utilization of its. . . features." This, Cohen and Cohen suggest, is further evidence that home video has not revolutionized the communication process.

Imperfectly knowledgeable of how to make their VCR work or not, many viewers do know enough to fast-forward through up to two-thirds of all commercials when they replay time-shifted programs. That's the major conclusion of a series of experiments done by Sapolsky and Forrest and reported in "Measuring VCR Ad-voidance." Judging from their experiments at least, Sapolsky and Forrest conclude that "zipping behavior among adults substantiates the concerns of the television industry."

Rounding out Part III is Roe's "School Achievement, Self-esteem and Adolescents' Video Use." Against the background of a highly charged policy debate in Sweden, Roe probes the relationship between doing well in school, feeling good about oneself, and VCR use. Roe finds weak and somewhat inconsistent relationships; but concludes that when legitimate achievement systems fail, students turn to violent and pornographic home video as substitutes for success and to participate in an esteem-bolstering youth subculture.

Roe's study on youth subcultures exemplifies an important strand of video research, which has focused particularly on the consequences of video use for children and adolescents (for earlier examples, see, Barker, 1984; Barlow & Hill, 1985; Roe, 1985; and Greenberg &

Heeter, 1987). Future research could extend this line of inquiry, first to other kinds of content and its effects on children (e.g., learning of social values or acquisition of cognitive skills); and second, to the adult VCR users, where one might investigate, for example, the consequences for adult information levels and behaviors of viewing, say, prerecorded how-to-do-it tapes, compilations of news for specialized audiences, or shopping catalogues. Similar questions could be asked about the possible effects of time-shifted content such as TV dramas, soap operas, or sports.

In both instances, researchers might reasonably expect to find at least some stronger effects from video use than from television exposure, since VCR use implies greater interest in the content, possibly higher levels of attention while viewing, and potentially greater amounts of background knowledge, which would aid in recall and comprehension.

While focusing on the effects of VCR-mediated *content*, investigators should not overlook the possibility that there are process-centered effects associated with video as well. It has been suggested, for example, that VCR use may be characterized by higher levels of audience activity than traditional television viewing (Levy, 1987). If VCR users have become a more selective and involved audience because of their video use, then it becomes interesting to speculate whether that heightened activity carries over to other mass communication experiences. Does having a VCR, for example, lead to greater selectivity in exposure to off-air television? In other words, are video users somehow changed from being individuals who watch television, any television, into consumers of specifically chosen programs?

VCRS, GROUPS, AND SOCIETIES

On a more philosophical note, home video, like the older technologies of mass communication, has become the subject of considerable speculation with regard to its impact on the quality of the human condition. The underlying issue here is whether the media in general and video in particular exert a centrifugal or centripetal force on social organization and the place of the individual in it.

Closely related to the question of social integration is the issue of cultural diversity. One of the great promises of VCR technology has been that it will substantially increase the variety of media content available to audiences. The VCR household, it is said, becomes its own television programmer, picking and choosing, from available messages, *what* it watches and *when*. Indeed, this promise is one of the key assumptions underlying policy making by the Reaganite F.C.C.

How much of this promise of media diversity is being met—and for whom—remains largely unexamined. However, at least one early study (Levy, 1980) found that VCR users tended to specialize in a relatively small number of existing television genres (e.g., soap operas, movies, children's programs) and that VCR audiences were simply exposed to more of the same, albeit self-selected, types of content. This finding raises the possibility that control over content, a significant part of the public definition of video, may be an illusion and that video does little to promote cultural pluralism.

Certain aspects of video seem ready-made for study along these lines. Across all levels of social analysis—individual, group, collectivity, societal—research could be framed which asks first, "How is video used?"; and second and more importantly, "What are the consequences of those use patterns for social cohesion?"

The final four chapters explore some of these questions. In "Away from the Mainstream: VCRs and Ethnic Identity," Dobrow demonstrates that some immigrant groups to the United States use home video almost exclusively to view foreign language cassettes. Tapes of movies and televised entertainment and sporting events were watched, often in groups, in order to feel connected with the old country. The role of home video in maintaining those cultural ties, Dobrow suggests, is an important question for scholars of ethnic identity.

Writing in "VCR Narrowcasting in the Kibbutz," Shinar examines the introduction of home video to the Israeli collective rural communities, known as *kibbutzim*. Drawing on models of democratic access, participation, and self-management, Shinar judges home video (and the closely related technology of kibbutz cable TV) to be a failure. Even in the ideologically active community of the kibbutz, VCR technology, Shinar concludes, has been unable to demystify or reinvent the television medium.

Shifting from the group to the national level of analysis, Ogan ("The Worldwide Cultural and Economic Impact of Video") provides data on VCR penetration in 67 countries; and examines the impact of home video and tape piracy on worldwide film production. As to its cultural effects, the VCR, Ogan suggests, is a "liberating technology that allows viewers a range of taste in media content never before possible. . . ."

The kind of previously unavailable media content viewed is one of the issues investigated in Boyd's "The Videocassette Recorder in the USSR and Soviet-Bloc Countries." Soviet citizens, Boyd reports, acquire home video to make money in an underground business activity, to learn about the West from non-Soviet media, and to engage in political activity. Although Boyd doubts that the VCR in and of itself will bring about major changes in the Soviet Union, he concludes that home video at least "opens the door to previously unavailable information."

Conclusion

This chapter and the book it introduces offer an overview of the VCR phenomenon and point to some of what is known and not yet known about video. By my choice, *The VCR Age* did not attempt to cover all VCR-related research. Among the interesting and important questions not discussed include how home video is used in the worlds of business, education, and politics. How, for example, do businesses use video to supplement inter-office memos? Why are Presidential candidates preparing pretaped cassettes for showing at fund-raising coffee klatches? How do political action organizations ranging from the AFL-CIO to the Environmental Defense Fund use video to recruit and retain new members?

This gap raises significant issues not only about the scope of this volume, but, more importantly, about fundamental definitions of mass communication and its study. The examples cited above, and indeed several of the chapters collected here, suggest that interesting questions that can be asked about home video are those which span traditional notions of interpersonal, mass, and organizational communication. Therefore, let me conclude by returning to the title of this chapter. Why aren't VCRs like pop-up toasters? Because VCRs don't make toast. What they do is make for an unexplored new set of problems and issues and a new challenge for communication scholars.

REFERENCES

Baboulin, J., Gaudin, J. & Mallein, P. (1983). *Le Magnetoscope au Quotidien: Un demi-pouce de liberte*. Paris: Aubier Montaigne.

Barker, M. (Ed.) (1984). *Video nasties: Freedom and censorship in the media*. London: Pluto Press.
Barlow, G. & A. Hill (Eds.) (1985). *Video violence and children*. London: Hodder & Stoughton.
Boyd, D. (1987). Home video diffusion and utilization in the Arabian gulf states. *American Behavioral Scientist*, *30*(5), 544-554.
Buckwalter, L. (1978). *The complete home video book*. New York: Bantam.
Cohen, A., M. Levy & K. Golden (1987). *Children's uses and gratifications of home VCRs: Evolution or revolution*. Presented to the annual meeting of the International Communication Association, Montreal, May.
Cubbitt, S. (1986). *Time shift: The specificity of video viewing*. Paper presented at the Television Studies Conference, London.
Darkow, M. (1985). Video in the Federal Republic of Germany. *EBU Review (Programmes, Law, Administration)*, *35*(4), 26-28.
Greenberg, B. & Heeter, C. (1987). VCRs and young people. *American Behavioral Scientist*, *30*(5), 509-521.
Gubern, R. (1985). La antropotronica: Nuevos modelos tecnoculturales de la sociedad mass-mediatica. In R. Rispa (Ed.), *Nuevas tecnologias en la vida cultural Española*. Madrid: FUNDESCO.
Kelly, R. (1988). *The economics of special interest video*. Presented to the Aspen Institute Conference, The Future in Home Video for Cultural Niche Programming. Wye Woods, Maryland, November 9-11.
Levy, M. (1980). Program playback preferences in VCR households. *Journal of Broadcasting & Electronic Media*, *24*, 327-336.
Levy, M. (1983). The time-shifting use of home video recorders. *Journal of Broadcasting & Electronic Media*, *27*, 263-268.
Levy, M. (1987). VCR use and the concept of audience activity. *Communication Quarterly*, *35*, 267-275.
Levy, M. & Gunter, B. (1988). *Home video and the changing nature of the audience*. London and Paris: John Libby.
McQuail, D. (1986). Is media theory adequate to the challenge of new communication technologies? In M. Ferguson (Ed.), *New communication technologies and the public interest*, (pp. 1-17). London: Sage.
McQuail, D. (1987). *Mass communication theory: An introduction*. (Second Edition). London: Sage.
Roe, K. (1985) The Swedish moral panic over video: 1980-1984. *The NORDICOM Review of Nordic Mass Communication Research*, 20-25, June.
Roe, K. (1987) Adolescents' video use. *American Behavioral Scientist*, *30*(5), 522-532.
Rubin, A. & Bantz, C. (1987). Utility of videocassette recorders. *American Behavioral Scientist*, *30*(5), 471-485.
Schoenbach, K. & J. Hackforth (1987). Video in West German households. *American Behavioral Scientist*, *30*(5), 533-543.
Staff. (1980, October). Videocassette recorders. *Consumer Reports*, pp. 590-594.
Tydeman, J. & E. Kelm (1986). *New media in Europe*. London: McGraw-Hill.
Williams, F., Phillips, A. & Lum, P. (1985). Gratifications associated with new communication technologies. In K. Rosengren, Wenner, L. & Palmgreen, P. (Eds.), *Media gratifications research: Current perspectives* (pp. 241-254). Beverly Hills, CA: Sage.

Part I
The Growth of Home Video

2

The Diffusion of the VCR in the United States

BRUCE C. KLOPFENSTEIN

The diffusion of home video accelerated more rapidly than any of the new television technologies by the mid-1980s. Unlike other new communication technologies that have yet to live up to the hype of the early 1980s (e.g., videotex, teletext, direct broadcast satellites, pay-per-view, and interactive cable television), the diffusion of videocassette recorders (VCRs) in the United States occurred more rapidly than was predicted in the late 1970s (Klopfenstein, 1989; Klopfenstein, 1985). Fifty-two percent of U.S. television households had adopted videocassette recorders by the end of 1987 (Electronic Industries Association, 1988), and one survey indicated VCR penetration had reached 65% by spring 1988. Of those households, 8% already owned two or more and 2% owned three or more (*Television Digest*, 1988, June 27, pp. 13-14). Fairfield Group estimated that 18% of all VCR households had more than one VCR by fall 1988 (*Television Digest*, 1988, October 31, p. 14). In urban areas, both Arbitron and Nielsen figures indicated the VCR penetration was higher than the overall *national* averages. Color television remained the only consumer electronics technology with a U.S. household penetration higher than that of the VCR; even black-and-white TV penetration (58%) is now lower than VCR penetration (Electronic Industries Association, January 1989).

The market success of the VCR, however, was not anticipated only five years earlier. Most industry experts believed that the VCR would remain an expensive luxury item for a limited number of generally upper income households (Klopfenstein, 1985). Use of the VCR to record television programs seemed destined to be a limited application of the technology. Since movie studios initially regarded videocassettes as a potential threat rather than a new source of income, few foresaw the rapid growth of what has become a billion-dollar video rental industry.

This chapter examines the historical diffusion of the VCR in the United States from its technical beginnings in the 1950s, to its U.S. introduction in 1976, and on to the present. The recent diffusion of the VCR is put into the context of previous communication technologies. Expectations for the future diffusion of VCRs are then presented.

A BRIEF HISTORY OF VIDEOCASSETTE RECORDERS

The history of video recording technology dates back to the 1920s.[1] Table 2.1 summarizes key developments in the history of home video generally and the VCR in particular. The emphasis in the table is on technical developments in home video including those which occurred with video disc players. The table gives a chronology of important events in home video history through 1988. A perusal of the table indicates how quickly the VCR industry progressed from the late 1970s to the early 1980s.

Video tape was invented in the early 1950s for use by television stations and networks. Video tape equipment became continually smaller and less expensive in the 1960s, making a potential home market for the technology seem possible. A number of manufacturers unsuccessfully tried to crack the home market in the late 1960s and early 1970s with difficult-to-use open-reel video tape recorders (VTRs) (see Klopfenstein, 1985). Sony invented the cassette recorder with its professional U-Matic machine, which was introduced in 1972. The Sony Betamax was the first *home* video device to be adopted by consumers starting in 1975 with a combination television set and VCR.

In Japan, Matsushita's independent subsidiary, Japan Victor Corporation (JVC) was developing its own videocassette recorder for the home. Company officials believed that a 2-hour machine would be much more appealing to consumers than Sony's one-hour Betamax (Nayak & Ketteringham, 1986). JVC soon introduced its incompatible "Video Home System" (VHS) VCR, and within two years, the lower-

Table 2.1 Home Video Chronology of Events

1927 John Logie Baird uses waxed phonograph discs to record TV images in Britain.
1951 Bing Crosby demonstrates magnetic videotape recorder (VTR).
1956 Ampex videotape recorder introduced for television use.
1959 Toshiba and JVC introduce helical scanning VTR.
1961 Peter Goldmark of CBS Laboratories begins work on film cartridge playback system.
1962 First optical videodisc demonstrated at SRI International.
1963 Nieman Marcus catalog includes $30,000 Ampex VR-1500 VTR.
1964 Goldmark demonstrates Electronic Video Recorder (EVR) film cartridge playback device internally at CBS.
1965 Westinghouse reportedly can produce still pictures on a videodisc.
1966 Various manufacturers are working on open-reel VTRs for home use. Sony introduces CV-2000 VTR.
1967 Ampex said to have sold around 500 VTRs to home users. Carnegie Commission cites CBS's EVR as being less expensive than videotape for storing programs.
1968 Several early home VTRs are available at $800 to $4,000; all will fail due to high price and/or technical difficulties.
1969 Sony demonstrates first videocassette recorder (VCR).
1970 Teldec demonstrates its limited videodisc system in Europe and U.S. CBS introduces EVR.
1971 Kodak announces 8mm film cartridge for television sets.
1972 Sony markets 3/4" U-Matic VCRs for $1,600. CBS drops EVR at a loss of $25-30 million. MCA demonstrates its optical videodisc system. RCA is reported to be working on a videodisc player. Cartrivision 1/2" video system is launched but fails after 8,000 are sold.
1973 JVC accelerates its VCR development. MCA and Philips announce work to make their optical videodisc systems compatible.
1974 MCA and Philips announce joint optical videodisc venture.
1975 Sony introduces 1/2" Betamax VCR with 1-hour recording capability to the U.S. for $2,300 which includes television set. Teldec videodisc player marketed in Germany. RCA and MCA/Philips demonstrate incompatible video-disc systems.
1976 Sony Betamax stand-alone VCR sells for $1,300. Teldec is discontinued in Europe. Matsushita subsidiary Japan Victor (JVC) has developed incompatible VHS "Vidstar" VCR with longer recording capability.
1977 RCA markets VHS VCR with 2-hour recording capability at $1,000. Sony introduces 2-hour Betamax. RCA debuts 4-hour VHS.
1978 JVC announces slow-motion and stop-action features for VHS. VHS success surpasses expectations. RCA redesigns VDP. In December Philips' Magnavox test markets "Magnavision" optical videodisc in Atlanta at $695. Two hundred discs at $6-$16 are available.
1979 800,000 U.S. homes have VCRs by early in the year. VHS adds 6-hour cassette. Beta evolves into 3-hour cassette. Only 1 prerecorded tape is sold for every 2 VCRs. Magnavision VDP and disc prices are increased. RCA announces intention to market its videodisc player.
1980 GE announces it will back Matsushita's "VHD" videodisc system (a third format). Many Magnavision videodiscs on the market are faulty. The 8mm camcorder is demonstrated.

(continued)

Table 2.1 (continued)

1981 Ninth Circuit Court of Appeals finds Sony guilty of copyright infringement. For the first time more video cameras are sold than Super-8mm film cameras. RCA introduces "Selectavision" videodisc player in March at $500 with only 100 disc titles available at $15-28. Fewer are sold than expected.

1982 Price war forces Beta and VHS VCR prices down to near cost. RCA cuts price on videodisc player to $299; Zenith sells compatible videodisc player for $170. Only 200,000 RCA videodisc players have been sold.

1983 VCR sales exceed expectations as competition continues to keep prices lower and stimulate additional features. Beta Hi-Fi, VHS Hi-Fi and 8mm video are developed. First camcorder (Sony Betamovie Camcorder) is introduced. 1983 RCA videodisc player sales total 300,000. VCR sales total more than 3,000,000.

1984 The U.S. Supreme Court overturns the lower court ruling in *Universal* vs. *Sony* concluding that VCRs do not violate copyright. 8mm VCR is introduced. RCA to discontinue marketing its videodisc player. 8-hour VHS tape is available.

1985 Beta market share down to about 10% of VCRs sold.

1986 13 million VCRs are sold including 1 million camcorders. U.S. penetration reaches 40% of households, higher in urban area. Stereo reception VCRs are introduced.

1987 VCR deck sales level off as the VCR has diffused into the majority of U.S. households. Super-VHS is introduced.

1988 Sony announces it will begin marketing VHS VCRs, leading observers to conclude the Beta format is all-but-dead in the U.S. and Europe; Beta's share of the U.S. VCR market is less than 1% of current VCR sales. VCR deck sales decline for the first time while camcorder sales continue to accelerate. Sony markets Video Walkman 8mm portable VCR.

1989 Consumer electronics industry observers expect VCR sales to remain close to recent years. Some believe the laser videodisc player will soon become a factor in the home video market as amount of available software continues to increase and consumers learn of the high quality of video produced.

SOURCES: Klopfenstein (1985); *Television Digest* (1985, various issues); *Dealerscope Merchandising* (May, 1988, p. 30).

priced, longer-recording VHS controlled 57% of the U.S. VCR market (Nulty, 1979). *Total* VCR annual unit sales rose dramatically from 402,000 in 1978 to 3,354,000 in 1983, and a record 13.2 million in 1986 (Electronic Industries Association, 1988). As seen in Table 2.2, sales of VCR decks leveled off in 1987-1988 as market limits were neared. Deck sales in 1988 were down only 8% from 1987; in light of the dramatically shrinking market of first time VCR adopters, the 1988 sales figures are remarkable. Advanced and more expensive S-VHS (Super-VHS) and ED-Beta (Enhanced Definition) are expected to revitalize dollar value sales of VCRs into the 1990s.

Table 2.2 U.S. VCR Sales to Dealers, HH Penetration, and Average Retail Price: 1975-1989*

Year	EIA Units Sold	Nielsen Penetration Estimates	TV Digest Avg. Factory VCR Prices	Merchandising Average Retail Price/VCR Deck
1975	30	—	$862.30	$2300
1976	55	.1%	$921.84	$1200
1977	160	.2	$721.65	$1100
1978	402	.3	$785.15	$887
1979	475	.5	$796.68	$902
1980	805	1.1	$773.13	$871
1981	1,361	1.8	$766.13	$795
1982	2,035	3.1	$645.18	$662
1983	4,091[a]	5.5	$523.94	$573
1984	7,616	10.6	$454.89	$482
1985	11,853	20.8	$401.88	$450
1986	13,174	36.0	$388.53	$399
1987	13,306	48.7	$384*	$389
1988	12,792	58.0	$385*	$404**
1989	13,400*	65.0	na	$408**

*Estimated
**EIA Estimate
a. Includes portable and camcorders sales, 1983-1987; camcorder sales were 517,000 in 1985; 1,169,000 in 1986; 1,604,000 in 1987; 2,044,045 in 1988; and estimated 2,700,000 in 1989.
SOURCES: Electronics Industries Association (1987); Nielsen as cited in Television Bureau of Advertising (1988); *Television Digest* (1987, June 1, p. 12; 1989, January 2, p. 10, January 23, p. 11); *Dealerscope Merchandising* (annual March statistical reviews). 1988-1989 retail prices are derived from EIA factory price estimates.

VCR penetration grew rapidly from 1983-1987. The penetration figures listed in Table 2.2 are from Nielsen, which has been criticized by some in the advertising industry in the past for underestimating VCR presence in television households (e.g., Frank, 1986). For example, an Electronics Industries Association (EIA) commissioned study of 7,500 households found that VCR penetration was 35% at the end of 1985 (Zahradnik, 1986) while Nielsen lists its 1985 and 1986 figures as only 20.8% and 36.0% respectively. The spring 1988 poll cited above found 65% VCR penetration while Nielsen's February 1988 figure is 58.1%. One way to help correct for this apparent bias is to consider the oft-cited annual Nielsen figures to be as of January 1 of each respective year.[2]

Dollar sales of VCR units are expected to continue to increase into the 1990s as the sales of expensive camcorders (portable video cameras with a built-in VCR) begin to accelerate. By 1988, the Electronics Industries Association reported that sales of camcorders represented 16% of all VCR unit sales, up from 4% in 1985. As depicted visually in Figure 2.1, camcorder unit sales increased from 517 thousand in 1985 to over two million in 1988 and reached 7% household penetration at the start of 1989 (Electronics Industries Association, January 1989). More expensive S-VHS and ED Beta VCRs will also tend to keep dollar sales figures high.

A growing percentage of new VCR purchases were replacement VCRs for initial purchases. As noted at the outset, multi-VCR households were becoming more and more commonplace as 1990 approached. In fact, 1988 was the first year in which more than 50% of total VCR sales (including camcorders) were to homes that *already* had at least one VCR (*Television Digest*, January 4, p. 12). Approximately 45% of 1989 VCR sales were expected to be to first time VCR buyers (*Consumer Electronics*, 1988, January) which would put U.S. VCR penetration at around 70% by the end of 1989.

The VCR Versus the Videodisc Player

Lost in the historical development of the VCR market is the surprisingly important impact the videodisc player (VDP) had on VCR diffusion (Klopfenstein, 1985). While a home videodisc player for the mass market has yet to appear, efforts were made to crack this market in the early 1980s by two systems: one from the U.S.'s RCA and the other developed by Europe's Philips (see Table 2.1 and Graham, 1986). RCA promoted its videodisc player heavily when it was introduced in 1981. This not only generated interest in the RCA VDP, but heightened consumer awareness of home video generally. This is the first step in the adoption decision process for any innovation (Rogers, 1983; 1986).

The introduction of the VDP unexpectedly made the VCR a more attractive consumer electronics device. The Japanese VCR manufacturers responded to what they perceived as a potential threat to their home video dominance by lowering their VCR prices to what some observers believed were even lower than the cost of production (Anderson, 1982). While some consumers became interested in purchasing a VDP, many instead opted for the more versatile VCR as was indicated by the takeoff of VCR sales from 1981 to 1983 (Klopfenstein, 1985). The VCR could perform the playback function of the VDP and added

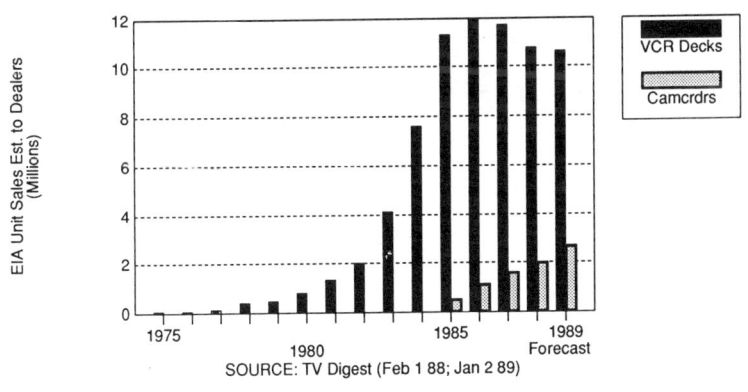

Figure 2.1. U.S. VCR/Camcorder Sales to Dealers

the recording function for about the same price as the Philips disc player.

Beta Versus VHS

From the point that the VHS VCR was introduced to the U.S. market by RCA in August 1977, the incompatible Beta and VHS VCRs battled for market dominance. It could be argued that it was really not much of a contest. JVC invented the VHS format, and their original goal was not to introduce a VCR until one had been developed which could record at least two hours of material. Sony's mission was to get to the market first and establish its Beta VCR as the technical standard (Nayak and Ketteringham, 1986).

Although VHS was introduced to the U.S. market after Beta had established a foothold, by 1979 it had already passed Beta in share of the VCR market. *Video Week* (August 25, 1986) reported that Beta's share dropped to 17% in 1984, 10% in 1985 and 5% in 1986. *Television Digest's* estimation of the loss of market share by Beta VCRs is visible in Figure 2.2.

Table 2.3 shows that dollar sales of Beta video tape peaked in 1984 and began to decline after that point. In 1987, 15 times as many blank

28 The Diffusion of the VCR in the United States

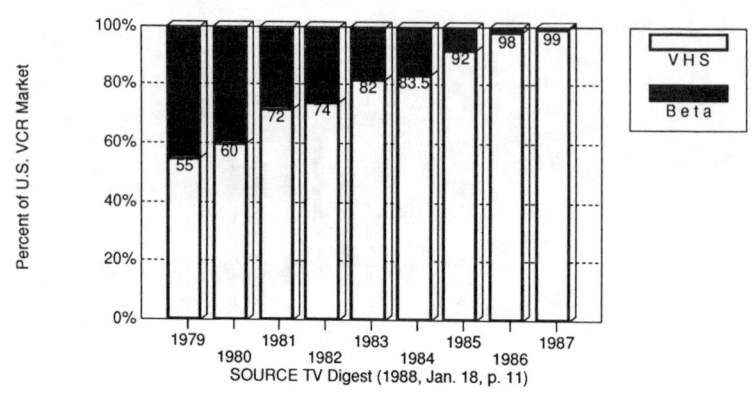

Figure 2.2. Growth of VHS Market Share in the U.S.

VHS tapes were sold as Beta. Many video stores stopped carrying the Beta format due to the much higher demand for VHS tapes. Table 2.4 shows that sales of *prerecorded* Beta tapes peaked in 1985 and have since declined. Sony announced in early 1988 that it would begin manufacturing VHS VCRs, a tacit admission that VHS had all but become the standard format. Fewer than 10% of VCR homes in 1988 were estimated to be Beta-only households.

JVC was correct in wanting to develop a VCR and cassette tape which could be used to record for a longer period of time. Beta's limited recording time versus VHS was a major factor in the format's dramatic loss of market share to VHS virtually from the start. By 1988, Beta tapes were limited to 5 1/2 hours of recording while VHS had a maximum of 8 hours (*Television Digest*, 1988, January 18, p. 11). VHS also held a critical price advantage, and as VHS market share continued to increase, so did the relative number of prerecorded titles available exclusively on VHS tapes. Although many held the perception that the Beta VCR produced a better picture than VHS, technical experts such as Weinstein (1984) and Prentis (1981) have concluded that this was, in fact, *not* the case; periodic reviews in *Consumer Reports* found VHS picture quality superior twice, found Beta superior once, and found no

Table 2.3 Blank Videotape Sales Estimates by Format, 1980-1987

	VHS		Beta	
Year	Value	Units	Value	Units
1980	$168,601	12,892	$63,699	6,142
1981	260,400	20,469	81,003	7,885
1982	310,043	27,429	102,299	11,201
1983	500,341	65,857	151,490	21,156
1984	714,405	122,012	199,281	35,931
1985	1,013,094	231,353	189,574	42,982
1986	1,250,095	312,785	137,267	34,171
1987	1,093,909	314,785	75,135	23,124

SOURCE: International Tape/Disc Association cited in *Television Digest* (1988, June 20, pp. 14-15). EIA estimates that a *total* of 365 million blank tapes (including 8mm were sold in 1988 with the same number expected in 1989. This is down from 393 million in 1987 (EIA, 1989).

Table 2.4 Prerecorded Videotape Sales Estimates by Format, 1980-1987

	VHS		Beta	
Year	Value	Units	Value	Units
1980	$139,932	2,028	$69,626	1,123
1981	268,204	3,529	111,724	1,643
1982	293,930	4,550	113,288	1,960
1983	389,163	6,973	131,492	2,633
1984	741,746	14,768	221,224	4,922
1985	1,612,641	35,670	300,618	7,430
1986	2,160,489	58,234	246,229	6,707
1987	2,210,828	62,701	196,983	5,947

SOURCE: Various *Dealerscope Merchandising* annual statistical reviews. These figures are based upon industry consensus derived by personal, mail, and telephone interviews with representatives of leading industry manufacturers. EIA estimates that 1988 sales of *all* prerecorded videotapes including 8mm increased to about $3 billion in 1988 with $3.25 billion expected in 1989. This is up from $2.8 billion in 1987 (EIA, 1989).

difference in a fourth review. In conclusion, the Beta format appeared to hold *no* advantages over the VHS other than being first on the market, and this may be a lesson for future marketers of new media products.

Price Drop

Table 2.5 breaks out the average price paid for the VCRs in the first half of the 1980s. These data show that more "low end" models were purchased as the decade progressed. Price was expected to be an important barrier to VCR diffusion in the minds of many industry forecasters, but VCR sales accelerated in 1979-1981 despite a small annual drop in average prices.

Two prices are listed in Table 2.2: one is an estimate of the factory price of all VCRs *including camcorders* (based upon EIA data) and another is the average *retail* price (based upon a survey of manufacturers). As seen in Table 2.2, prices dropped substantially from the previous year starting in 1982, although sales do not take off dramatically until about 1984. The RCA VDP introduction coincides with the drop in VCR prices.

In his study of various factors which influenced the takeoff of color television set sales in 1962, Beville (1966) concluded that price was only one of three factors (amount of color programming and obsolescence of black-and-white sets were the others) which were clearly correlated with the growth. Similarly, VCR growth could be tied to availability of software (prerecorded tapes as well as cable television). As will be seen, the VCR has diffused more rapidly than color television (although color TV was slowed as later adopters waited to replace their black-and-white sets).

The Growth of Prerecorded Cassettes and Video Rentals

The VCR had an advantage over other new media technologies. The "chicken and egg" software/hardware question, which is key to the diffusion of such new media as color television, compact discs, and even cable television, was not a major obstacle for VCR adopters. All broadcast and cable programs available were potential software sources. There was no inherent pull from VCR adopters for prerecorded cassettes. Nevertheless, entrepreneurs saw the opportunity VCRs presented to push prerecorded material (Lardner, 1988).

The growth of the prerecorded videocassette industry is a topic worthy of a book in and of itself. The growth of the prerecorded video market *followed* that of the VCR market. The majority of early prerecorded cassettes sold were sexually oriented "adult" titles. In both 1978 and 1979, for example, an estimated 75% of *all* prerecorded tapes

Table 2.5 Price Decline of VCRs

VCR Deck Retail Price Range	1981	1982	1983	1984	1985
Less than $300	7%	4%	7%	13%	22%
$300 - 499	22	32	53	53	56
$500 - 699	27	23	19	20	14
$700 and more	45	42	22	14	8

SOURCE: Electronic Industries Association as cited in *Discount Store News* (1986, June 9, p. 31).

sold were X-rated (*Merchandising*, 1980, March, p. 53). There are a number of possible reasons for this. Early on, there was very little prerecorded fare of any kind, and adult titles made up a large percentage of all titles available. While the major studios were reticent to put their material on cassette (Lardner, 1988), the smaller producers of adult material welcomed the new medium enthusiastically. The production cost of adult films was also far lower than mainstream films. Finally, early VCR adopters had private and, for the first time, ready access to intriguing, yet socially unacceptable material. As more movies on video became available, the adult material's dominance of prerecorded tapes was diluted. The percentage of prerecorded tapes sold which were X-rated dropped to 37% in 1980 and 33% in 1981 (*Merchandising*, 1982, March, p. 30).

Inexpensive videocassette rentals have dramatically increased the VCR's utility to users for playback of prerecorded movies. *Market Facts* projected that 40% of all films viewed in 1986 were already being seen at home on VCRs when VCR penetration was also around 40% (see Table 2.2). Teens and young adults (ages 20-29), the two most frequent movie-going groups, have decreased their theater attendance significantly (Roth, 1986). A more recent survey by the American Video Association indicated that the frequency of renting prerecorded tapes, however, was beginning to decline (Bierbaum, 1988). As VCR adopters rent (and perhaps copy) their favorite prerecorded video tapes, the utility of renting tapes may begin to decline. New releases may represent the only sustained attraction for future visits to the video store, although nonentertainment tapes may find more frequent use.

As seen in Table 2.4, the *growth* in the value of sales of prerecorded tapes lags well behind the increase in VCR penetration seen in Table 2.2 from 1984-1987. There are two explanations for this. Rentals became much more readily available to VCR owners during that time period, making purchase of prerecorded tapes less appealing. Secondly, the retail prices of prerecorded tapes also declined during this period.

Summary of Factors in the Growth of the VCR

Based on an extensive critical analysis of published information available at the time, Klopfenstein (1985) found a number of stimuli to the accelerating diffusion of VCRs. The format battle between Beta and VHS actually served as a catalyst to VCR diffusion. While the lack of a standard VCR format may have been confusing and may have slowed early consumer adoptions, two competing formats clearly were *not* a barrier later. Inter- (Beta versus VHS) and intra-format (e.g., competition between different VHS manufacturers) VCR competition as well as the perceived competitive threat from the VDP led to technological innovations. These advances included longer recording capabilities, programmability, stereo sound, lower manufacturing costs, and lower prices. The fierce competition also increased consumer awareness (the first step in the adoption process) via heavy advertising and sales promotions.

The primary use of the VCR by innovators and early adopters was for time-shifting broadcast programming, not for viewing prerecorded fare (e.g., Levy, 1981 and Levy, 1983). Sony advertised its VCR on this basis and was not interested in pursuing a market for prerecorded tapes. Thus, prerecorded software availability, which was initially dominated by adult material, does *not* appear to have been a key factor in early VCR adoptions. As prerecorded videos became widely available by the mid-1980s, however, the nonrecording function of the VCR did become a primary reason for VCR adoption by later adopters (Klopfenstein, 1988).

VCRs and Other Television Technologies

Klopfenstein (1985) discovered that the growth of color television is often used as a bench mark by which to compare the diffusion of contemporary communication technologies. More recently, observers have noted the similarity in the growth of VCRs to that of color television. Clearly one measure of the attractiveness of the VCR as a *new* communication technology to adopters would be a comparison of

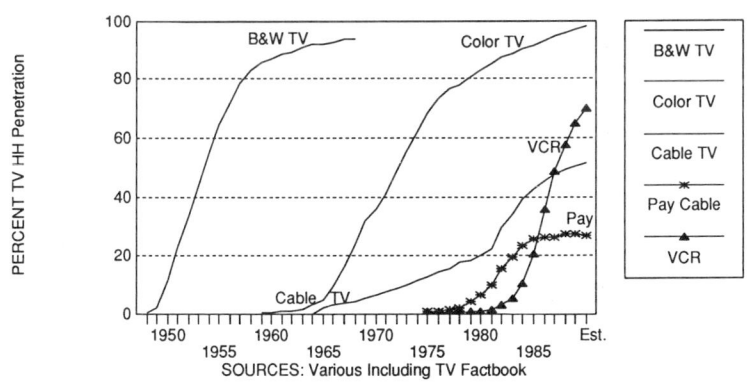

Figure 2.3. Relative Diffusion of Five Television Technologies

the relative diffusion rates of the VCR versus other past technologies (Carey and Moss, 1985, previously looked at the growth rates of various new media).

All new media technologies initially diffuse very slowly as the technology is perfected and prices are quite high. In addition, marketing research has repeatedly shown that a small number of innovators will adopt virtually *any* new technology or product. Thus, it is difficult to come to conclusions about the initial diffusion of the technologies based upon adoption by innovators. It is more instructive to examine how quickly the growth of the new technology accelerates once it has passed its nascent stage.

Figures 2.3 and 2.4 compare the growth of VCR household penetration to that of black-and-white, color, cable, and pay cable television. The lines are based upon penetration of television households except black-and-white television, which has penetration of all households. (While disaggregated annual black-and-white penetration data were not readily available past 1970, as noted earlier its household penetration has been *declining*; annual sales have decreased every year in the 1980s). Once VCR sales took off in the 1980s, the growth in household penetration is very comparable to that achieved by black-and-white television in the 1950s.

As seen in Figure 2.3, VCRs grew much more rapidly than cable. This is not surprising as there were limits on the diffusion of cable

34 *The Diffusion of the VCR in the United States*

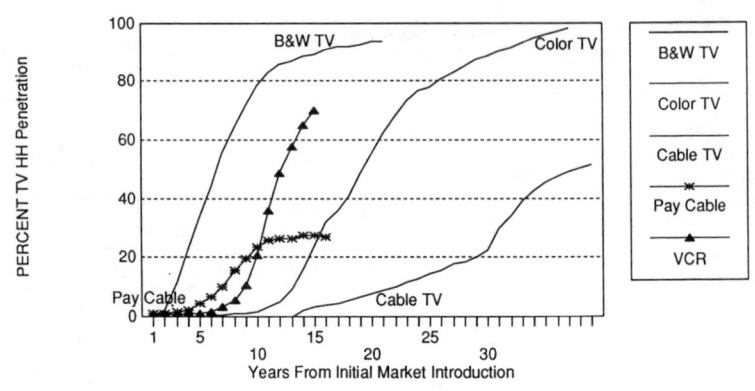

Figure 2.4. Superimposed Diffusion of Five Television Technologies

including the franchising process, satellite-delivered cable channels, and growth of alternative video services. Pay cable leveled off in the late 1980s and was expected to grow very slowly, if at all, into the early 1990s (Sylvester, 1988). Many see the VCR as an obstacle to further growth by pay-cable movie services; Klopfenstein (1988), for example, found that pay cable subscribers said they would prefer having a VCR to watch rented tapes if forced to choose between the VCR and a pay cable service.

Figure 2.4 allows a more direct comparison of the growth rates of the new television media. Each technology's growth curve is superimposed on the other. Once again in comparing VCRs to color television, Figure 2.4 reveals that VCR diffusion initially lagged behind that of black-and-white television, yet has clearly diffused more rapidly than color television. Color television's first year of diffusion is considered to be 1954, and the VCR's first year is 1975. Color television grew very slowly in the 1950s (Beville, 1966) and annual sales actually fluctuated each year in the 1960s *and* 1970s. Annual VCR sales in the 1980s have grown considerably more rapidly than annual color television sales in both the 1960s and 1970s (see *Television Digest*, 1988, August 8, p. 12). It took only ten years for VCR annual sales to reach ten million units (in 1986) while it took *25 years* for color television to do so (in 1979). Not only did the VCR take off sooner than color television, but Figure 2.4 reveals that the slope of its penetration curve is steeper than

Table 2.6 VCR Growth vs. Other Media Technologies

	Year Household Penetration Began	Year 30% Penetration Reached	Number of Years to Reach 30%
Radio	1923	1925-1930	Approx. 5
Television	1949	1950-1955	Approx. 5
Color TV	1954-1958	1969	12-15
Cable TV	1952	1982	30
Pay Cable	1975	1992	17
VCR	1975	1985	11

SOURCE: Adapted from Bailey, 1986, p. 9. Pay cable forecast of 30% penetration by 1992 source is Veronis, Suhler & Associates as cited in *Video Week* (1988, June 27, p. 5).

that of color television. This would seem to suggest that future VCR diffusion may continue at a rate *at least* equal to that of color television in the late 1970s.

Finally, Table 2.6 gives another way to gauge the diffusion of the VCR vis-à-vis other new media technologies. Radio was the first electronic medium, and it diffused quite rapidly in the 1920s (although reliable data were not kept at the time). Television was delayed by World War II, but consumers knew it was coming by the time it took off in the early 1950s. Although the VCR took 11 years to reach 30% penetration, it took color television considerably longer. *Only radio and television reached 30% of homes faster than VCRs.* Historically speaking, the VCR has diffused very rapidly in the U.S.

Future VCR Diffusion and Technologies

Clearly VCR unit sales *growth* reached a plateau by the late 1980s. The number of households without a VCR continues to decline, so the remaining market of first adopters continues to shrink. A look at past growth rates of any new media technology reveals the limits of growth are indicated by an inflection point in the growth curve. This point was reached by the VCR around 1986. The market for camcorders, however, is only beginning to be penetrated, as shown in Table 2.2.

In terms of the television audience, however, the demographic groups that do not yet have VCRs are generally the least attractive to most advertisers; conversely, those households with VCRs (including multiple-VCR households) are more socioeconomically upscale households, especially headed by persons in the 35-54 age group. Few broadcast television programs are targeted at very young adults or the elderly, the two households also least likely to be VCR households. Thus, the finite nature of the VCR market at this point may be less important to the broadcast industry than the behavior of those demographically attractive households which already have VCRs.

In terms of future VCR growth, diffusion theory predicts that it will progress more slowly among the relatively few remaining nonadopting households. This expected slowing in the diffusion rate will prompt some to conclude that the home video boom has ended (e.g., Farhi, 1989). Swanson and Klopfenstein (1987) forecast VCR penetration to 1990 using a demographic approach. Although both the youngest and oldest households will have the smallest VCR penetration, 90% of households headed by 40-59-year-olds will have VCRs *by 1990*.

Even if VCR prices remain relatively constant or even increase, the VCR penetration among households with younger heads will continue to increase. As established VCR households become multi-VCR households, two things may contribute to the diffusion of VCRs in younger households: (1) more of the young adults will come from VCR households, and may not consider the VCR a "luxury" item (a Gallup survey in spring 1988 found that 78% of all households with children already had a VCR); and (2) those young householders will receive new VCRs as gifts or may inherit the older VCRs from their nest households.

Households that already have VCRs will have more options from which to choose. Digital, multihead, and Multichannel Television Sound (MTS) VCRs are among the more exotic features which earlier U.S. VCR adopters may seek in second or third VCR purchases. MTS VCRs have grown from about 10% of the annual VCR sales market in 1986 to about 25% in 1989 (Electronic Industries Association, 1989). New Super VHS (S-VHS) VCRs and camcorders are already having a significant impact in Japan. Twenty percent of Japanese VCR sales in the first half of 1988 were S-VHS and 30% of camcorders were S-VHS. The U.S. has historically lagged about one year behind the Japanese market. Another format war may be on hand as Sony continues to push its nearly professional ED Beta VCRs, which produce 500 lines of resolution, better than S-VHS and better than nearly all television receivers today can display (*Television Digest*, 1988, August 15). These new VCR choices suggest that VCR diffusion may again accelerate at some point in the future, even as color television sales continue to set records. *Television Digest* (1989, January 2) expected the upturn to begin in 1989.

Another option is the new portable Sony Video Walkman, which includes a VCR in a package about the size of a large book (eight inches long). The video Walkman is a combination 8mm VCR and three-inch liquid crystal display (LCD) color TV (*Television Digest*, 1988, June 6, p. 16). While this product may not be widely adopted, it confirms that a trend toward further miniaturization and portability is continuing. Sony also continues to market its VHS-*incompatible* 8mm camcorders. There are already indications that the optical videodisc player many be poised to reemerge in the 1990s with its advantages of less expensive software, high quality video, and random access capabilities. Combination players that can play both videodiscs and compact discs (CDs) also may spur a VDP resurgence (Simels, 1988). Other looming digital technologies including erasable CDs and digital movies-on-demand delivered by telephone ensure that recording video will clearly be more commonplace in the household of the future than it is today.

VCRs diffused more rapidly in the 1980s than most experts predicted and have continued to outpace many expectations (e.g., Lancaster and Wright, 1983). While color television has often been used as the analogy for the growth of VCRs, perhaps future VCR sales could be viewed as more analogous to that of black-and-white television. Just as color television slowly replaced black-and-white, the advanced VCR systems will replace or be added to existing VCR(s) in the household. Television penetrated 55% of American homes in 1954, and it took eight more years to reach 90% penetration later in 1980. This would appear to be the most optimistic scenario for future VCR diffusion and could put VCR diffusion at 90% of American households as early as 1996. Although this forecast is more optimistic than virtually any available, the VCR has outpaced color television growth in the last few years. The growth in the number of older households could, however, keep VCR penetration from reaching 90% until closer to the turn of the century.

The number of multi-VCR households will also increase steadily. Various surveys of VCR adopters have shown a very high degree of satisfaction with the device (e.g., Klopfenstein, 1988). Although the VCR of the late 1980s may be replaced by higher resolution models and then digital models in the 1990s, video recording technology is becoming an integral part of the U.S. television household.

NOTES

1. A number of additional references give further details of the historical and technical developments of the VCR (Fox, 1983; Rosenbloom & Freeze, 1985; Shiraishi, 1985; Klopfenstein, 1985; Sugaya, 1986; Bailey, 1986; Graham, 1986; and Lardner,1987).

2. While the Nielsen figures are among the most conservative available for VCR penetration, it should be understood that exact VCR data are *not* available. *Television Digest* (1988, February 1; July 4) noted the discrepancies between Nielsen, Arbitron, EIA and U.S. Department of Commerce data. *Television Digest* derives some of its own data from EIA. The retail trade journal *Merchandising* makes annual VCR sales and price estimates based upon surveys of leading manufacturers. Nevertheless, the data in this chapter are the most commonly cited public information available.

REFERENCES

Anderson, T. W. (1982, January 13). Jury is still out on videodiscs. *Variety*, p. 38.

Bailey, J. D. (1986). Emergence of the home video market. In E. A. Lazer (Ed.), *Guide to videotape publishing* (pp. 1-19). White Plains, NY: Knowledge Industry Publications.

Beville, H. M., Jr. (1966). *The product life cycle theory applied to color television*. Unpublished masters thesis, New York University, New York.

Bierbaum, T. (1988, July 6). Survey backs video erosion; Rentals off, VCR resistance up. *Variety*, p. 35.

Carey, J., & Moss, M. (1985). The diffusion of new telecommunication technologies. *Telecommunication Policy 9*(2), 145-158.

Consumer electronics annual review. (1988). Washington, DC: Electronic Industries Association.

Consumer electronics U. S. Sales. (1989, January). Washington, DC: Electronic Industries Association

Farhi, P. (1989, January 14). Consumers hit pause button on sales of video recorders. *The Washington Post*, pp. A1, A11,

Fox, B. (1983). Videocassettes—Past, present and future. *Intermedia, 11* (4/5), 18-21.

Frank, B. (1986, October). When VCR meets NBC. *Marketing & Media Decisions*, pp. 106-110.

Graham, M. (1986). *RCA and the videodisc player: The business of research*. New York: Cambridge University Press.

Klopfenstein, B. C. (1989). Forecasting the market for home information services. *Journal of the American Society for Information Science, 40*(1), 17-26.

Klopfenstein, B. C. (1988, April) *The emerging VCR household: Relationships among ownership, demographics, and usage patterns*. Paper presented at the meeting of the Broadcast Education Association, Las Vegas, Nevada.

Klopfenstein, B. C. (1985). Forecasting the market for home video players: A retrospective analysis. *Dissertation Abstracts International, 46*, 546A. (University Microfilms No. 85-10588).

Lancaster, G. A., & Wright, G. (1983). Forecasting the future of video using a diffusion model. *European Journal of Marketing 17*(2), 70-79.
Lardner, J. (1987). *Fast forward*. New York: W.W. Norton.
Lardner, J. (1988, October). Romancing the cassette. *Video* pp. 60-65.
Levy, M. R. (1981). Home video recorders and time shifting. *Journalism Quarterly, 58,* 401-405.
Levy, M. R. (1983). Time-shifting use of home video recorders. *Journal of Broadcasting, 27*(3), 263-268.
Nayak, P. R., & Ketteringham, J. M. (1986). *Breakthroughs!* New York: Rawson Associates.
Nulty, P. (1979, July 16). Matsushita takes the lead in video recorders. *Fortune,* pp. 110-112.
Prentis, S. (1981, August). VHS meets Beta. *Popular Electronics,* pp. 38-43.
Rensberger, B. (1987, April 13). Lessons of the VCR revolution. *The Washington Post,* pp. A1, A10.
Rogers, E. M. (1983). *The diffusion of innovations.* New York: Free Press.
Rogers, E. M. (1986). *Communication technology: The new media in society.* New York: Free Press.
Rosenbloom, R. S., & Cusumaanbo, M. A. (1987). Technological pioneering and competitive advantage: The birth of the VCR industry. *California Management Review, 29*(4), 51-76.
Rosenbloom, R. S., & Freeze, K. J. (1985). Ampex corporation and video innovation. In R. S. Rosenbloom (Ed.), *Research on technological innovation, management, and policy,* (Vol. 2, pp. 113-185). Greenwich, CT: JAI Press.
Shiraishi, Y. (1985, December). History of home video recorder development. *SMPTE Journal 94*(12), 1257-1263.
Simels, S. (1988, November). Laser discs: Video's 10-year overnight sensation. *Video Review,* 44-46, 128.
Sylvester, A. (1988, September). Television-audience measurement in transition. *Marketing & Media Decisions,* 84, 88.
Sugaya, H. (1986, March). The videotape recorder: Its evolution and the present state of the art. *SMPTE Journal, 95*(3), 301-308.
Swanson, D., & Klopfenstein, B. (1987, December). How to forecast VCR penetration. *American Demographics,* pp. 44-45.
The videocassette recorder. An American love affair. (1985). Washington, DC: Television Digest.
Trends in television. (1988). New York: Television Bureau of Advertising.
Trost, M. (1986, January 9). VCR sales explosion shakes up industry. *Advertising Age,* p. 14.
Weinstein, S. B. (1984). *Getting the Picture*. New York: IEEE Press.
Zahradnik, R. (1986, June). Rewinding VCR penetration. *Marketing & Media Decisions,* p. 28.

3

Home Video: The Consumer Impact

PAUL B. LINDSTROM

If one listens to the consumer press it would be assumed that the VCR phenomenon hit the media industry like the monstrous dragon of storybook lore. Many stories have predicted videophiles running roughshod over the television business as we know it today. The myth has it that home video will destroy pay television with the prior release of movies, bankrupt the networks who rely on commercial advertisements that are now being zapped, and create a generation of couch potatoes who will no longer be interested in going outside to movies for entertainment and who are ready only to absorb information from video.

Fortunately in my judgment, however, the myth of the VCR dragon far outstrips the reality. In this chapter, I will attempt to debunk some of the fable surrounding home video and to create a realistic framework for analyzing the impact of VCRs on the lives of American consumers. In order to accomplish this, I will examine three areas:

(1) Current statistics on VCR ownership;
(2) Patterns of VCR recording and playback within the context of overall television usage; and
(3) Rental of prerecorded video.

VCR OWNERSHIP

Nielsen Media Research has tracked VCR penetration in the Nielsen Station Index (NSI) diary sample since 1982. NSI measures television

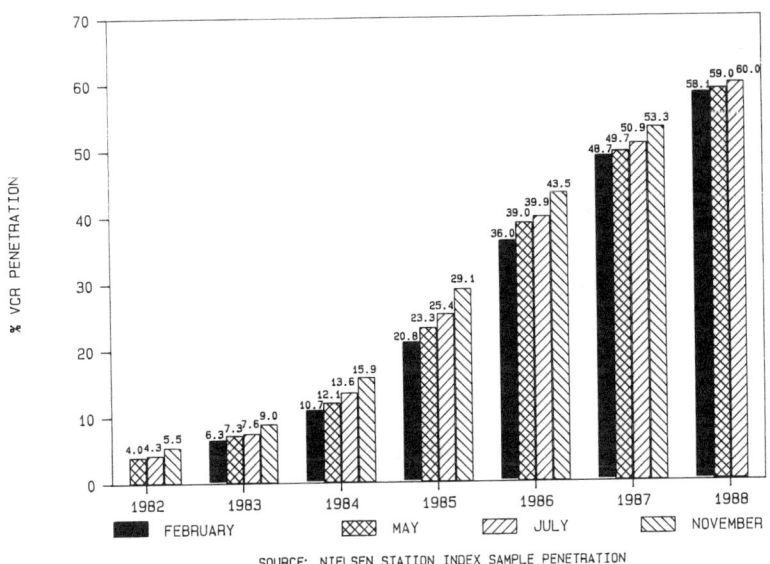

Figure 3.1. Growth in VCR Penetration

viewing in all markets across the country four times a year (February, May, July, and November). The sample size in each measurement is approximately 100,000 households. In the ten-year period since the VCRs first major consumer push, penetration has risen steadily, reaching a level of 60.0% ownership among all television households by July, 1988.

Ownership of VCRs is concentrated in large urban areas, particularly on the east and west coasts. In November, 1987, for example, VCR penetration was 57.2% in the NTI northeast region, 60.8% in the NTI pacific region, but only 49.4% in the east central and 49.1% in the south. The urban skew of video ownership is demonstrated most notably by the rapid decrease in penetration when ownership is examined by the size of the television market. Ownership levels (November, 1987) declined from 60.7% of TV households in the top ten markets to 52.2% in markets ranked 11-20, to 47.4% among markets 101+. Interestingly, the urban skew of VCRs is opposite to that found for cable TV, also an

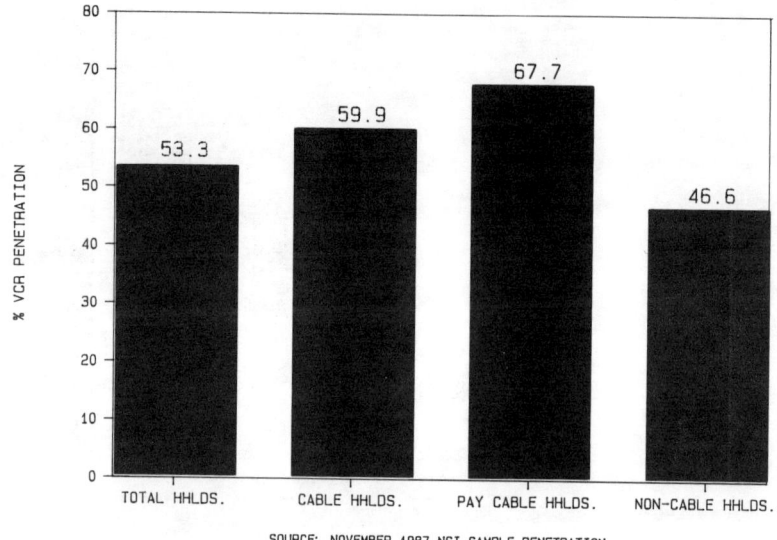

Figure 3.2. VCR Penetration by Cable Status

alternative video delivery system; cable penetration is a phenomenon of the smaller, less urban areas.

Since VCR ownership tends to skew towards urban households and cable skews toward rural homes one would not expect a positive correlation between the two. However VCR penetration is higher than average among cable households. The cable penetration figure is misleading though. In November, 1987, VCR penetration was 67.7% among pay cable homes. Among households taking only basic cable service, VCR penetration was 52.1%, a figure more comparable to noncable homes. It is therefore the high penetration level among *paycable* households that raises the total cable penetration to higher than average levels.

Pay cable penetration, like VCR penetration, also tends to skew urban. Pay cable and VCR households are similar in makeup, because the home video and pay cable products, although marketed differently, are the same to the consumer: uncut movies in your home.

Although both pay cable and VCRs appeal to the same market, they do not necessarily cannibalize one another. Most of the evidence I have seen suggests that VCRs and pay cable supplement or complement one another. For example, among VCR households, pay cable penetration declined from 37% to 36% from November 1984 to November 1987; while among the entire U.S. population, pay cable penetration increased from 26% to 28% during the same time period. However, the number of homes having both a VCR and pay cable actually increased by almost 250%.

This phenomenon illustrates the complexities of examining trends or changes within an ever increasing household base. Since pay cable homes and VCR households are similar in nature, the first homes to buy VCRs would most likely have been those making heavy use of the pay cable. One would thus expect that as the VCR base broadened, the percentage of homes with pay cable *should* decrease. Therefore, the most surprising findings are not that the percentage of homes subscribing to pay cable has decreased, but rather that the proportion of pay cable subscriptions has remainded basically unchanged.

Pay penetration is not the only characteristic of VCR homes that has remained stable. From the time that the first Nielsen Homevideo Index (NHI) VCR studies were conducted in 1982, when VCR penetration was 4%, up to the present 58% penetration, many VCR household characteristics have shown little change. The fact that many of the VCR demographics have not changed indicates that we may be reaching a penetration plateau and raises the question of how high VCR penetration can rise. A lack of change in household characteristics by definition means a finite market.

Let's examine a few key areas. In terms of composition, we find that since 1985 the VCR households have become somewhat older and there has been a sizeable decrease in the number of teens and children. The greatest number of composition changes occurred between 1985 and 1986. Since then the household composition has remained stable. There has been, for example, a 2% decrease in household size since fourth quarter 1986. VCR households are slightly larger (264 persons per 100 households) than U.S. television households in general (259 persons per 100 households).

The percentage of male heads-of-house with a college degree education or better in VCR homes has remained stable since 1985 at approximately 41%. This compares to the national figure of 21%. The percentage of VCR homes where male heads-of-house have professional or managerial positions has changed from 48% to 43%. This represents a small decrease over the last three years, but VCR homes remain well above national average on this measure.

In terms of income, NHI's VCR usage studies indicate that in 1982, 87% of VCR households had an income of $20,000 or more and 59% had an income of $35,000 or more. During 1987, 88% of VCR households had incomes of $20,000 or more, and 57% had incomes at or above $35,000.

Despite the overall lower prices of VCR hardware, software, and blank tapes, VCRs remain an upscale product. The VCR penetration level among households with less than $20,000 income was only 16.8% in November 1987. This figure rises to 73% among households with $20,000 to $50,000 incomes and to 82.7% among those with $50,000+ incomes.

This data may indicate the extent of the finite VCR market. VCR ownership has already reached near saturation level among upper income homes. It will require a major change in purchasing patterns among lower households for growth in VCR penetration to continue.

VCR Use in the General Television Environment

While there have not been major changes over the years in the demographic characteristics of VCR households, there are significant differences between the media behaviors of earlier and more recent owners. First of all, the earliest adopters of VCRs are consistently much heavier viewers of television in general. In the NSI diaries for May, 1987, for example, households which had a VCR for less than 1 year watched 145 quarter-hours of television. By contrast, households which had a VCR for 1-2 years watched 154 quarter-hours of television, those which had their VCR for 2-3 years watched 155 quarter-hours of television, and those households owning a VCR for 3 years or more years viewed 162 quarter-hours of TV.

In general, recording activity follows TV activity, with lighter television usage roughly translating into lighter VCR recording activity. The average household owning their machine for less than 1 year made 7.1 recordings during the average month of fourth quarter 1987. This compares to 9.3 recordings for those owning 1-2 years, 9.8 recordings for those owning 2-3 years, and 11.3 recordings in households owning a VCR for 3 or more years.

Overall, VCR households made 10 recordings during the average month of the fourth quarter of 1987. However, it is important to keep in mind that VCR activity follows the same 80/20 rule that many consumer goods do. Nielsen defines households that made 21 or more recordings during the average month as heavy recorders, and the 16% of VCR households defined as heavy recorders accounted for 55% of all recordings. When these heavy recorders are excluded from the averages, one finds that the average VCR household made only 5.3 recordings during the month or a little more than one recording per week. These 5.3 recordings per month for 84% of the VCR households can be compared to approximately 164 hours of television usage during the month for those same households. Thus, video use is only a small proportion of total television use.

There is yet another interesting point regarding the heavy recorder phenomenon. In 66% of the heavy recording households the primary VCR user was female, with men accounting for 30% of heavy use and nonadults, 4%. Thus, the majority of VCR recording was done by women 18 years of age or older.

Overall among all VCR households the average minutes of recording usage per week has declined since 1985, while the average minutes of playback has remained approximately the same. The Nielsen Television Index VCR Status Report (1985-1987) found an average 171 minutes per week of VCR recording and 253 minutes of VCR playback per week in January-December, 1985. Similar averages for January-December, 1986 were 172 minutes of recording and 273 minutes of playback; while for January-December, 1987, comparable figures were 154 minutes of recording and 258 minutes of VCR playback.

The decline in recording is a function of the influx of lighter television users and, thus, lighter recorders among the newest VCR owners. Playback usage reached its peak in 1986. This is a reflection of the growth and subsequent plateau of rental behavior. (See "Rental activity" which follows). The decrease in playback of 15 minutes per week corresponds to the 18 minute per week decrease in recording activity which has occured since 1984.

The question remains: How much impact does the VCR recording have on the audience for network programs? So far, the answer is relatively little. The average prime time program during May 1987 had less than 2.5% of its audience accounted for by unattended recording, that is, by households that recorded a program and either did not view television or viewed a different program while the recording was made.

RENTAL ACTIVITY

The home video industry witnessed a phenomenal growth in the 1984-1988 period. According to Nielsen's quarterly surveys of VCR usage, the percentage of VCR households that have rented prerecorded tapes "in the last year" has risen from 49% in 1982, to 80% in 1985, and to 90% in 1986, where it has remained relatively constant.

The growth in number of *households* renting prerecorded cassettes and in the number of *cassettes* rented has been greater than the increase in the number of VCR households. This is due to a rise in the percentage of households that rented during the average month. In the third quarter of 1984, for example, 40% of VCR households rented tapes. The figure rose to 49% during the fourth quarter 1987. The growth of the number of renting households and rental volume "follow" each other as the number of tapes rented per month by households that rent has changed little since 1984.

Rental dollar volume growth paralleled VCR household growth from the third quarter 1984 to the first quarter 1986, despite the rise in the number of renting households. This was due to a gradual erosion of rental prices. The average price for a night's rental during the third quarter 1984 was $2.61. The cost dropped to $2.27 by the first quarter of 1986. Since that time the growth in dollar volume has fallen substantially below that of VCR households because of the continual erosion of rental fees. By the fourth quarter 1987, the average daily rental cost was $2.06.

The changes in the rental marketplace are a reflection of fundamental changes in consumer motivation as witnessed by their behavior. The initial VCR purchasers were enticed into the marketplace because they wanted more television. These earlier adopters were very active recorders. According to the NHI quarterly VCR report for the fourth quarter of 1984, the average VCR household made 15.4 recordings per month. By the fourth quarter of 1987, the average VCR household was making 10.0 recordings per month. These figures represent the influx of new VCR owners, consumers who were less interested in increasing their television exposure.

Indeed, in the period roughly 1984-85 there was a large influx of consumers who appeared to be enticed into purchasing a VCR because they wanted to be able to watch uncut, commercial-free motion pictures at home at their convenience. These newer owners rented at substantially higher rates than early adopters. During the third quarter 1984, for example, households owning a VCR less then a year averaged 3.1 rentals per month, compared with 2.4 rentals per month for households that owned for 1-2 years, 1.8 rentals per month in households that had owned for 2-3 years, and 1.4 rentals per month for households that had owned their VCR for 3 or more years. Similarly in the third quarter of 1985, households that owned a VCR for less than 1 year averaged 4.4 rentals per month, contrasted with an average of 2.8 rentals per month in 1-2 year old VCR households, 2.2 per month in 2-3 year old VCR households, and 3.0 in households that have owned a VCR for 3 or more years.

As the home video industry was in a period of exponential growth, these new homes had a major impact on overall averages, causing rental volume to increase faster than VCR penetration. After the cream of the rental crop was absorbed into the VCR mainstream, the only area of growth open to the home video industry was among consumers who turned out to be lighter, more normal use renters. In the third quarter of 1987, for example, all VCR households rented an average of 2.9 tapes per month. In the same period, households that had owned their VCR for either less than a year or 1-2 years rented only 3.1 tapes on average. In short, their rental behavior more closely approximated that of the VCR population as a whole. Therefore without the inflation caused by the abnormally heavy use by new owners, the average frequency of rental declined and growth in volume fell below that of VCR households.

THE MEASUREMENT OF PRERECORDED CASSETTES ENTERS THE ELECTRONIC AGE

Since 1986 the market for sales of nonsell-through titles has remained flat. During this period major motion pictures, priced at $79.95-$99.95 at retail, have shipped approximately 400,000 units per title. This plateau is primarily the result of the saturation and subsequent halt to expansion of the number of new video stores. One store with 2,000 renters club members requires fewer copies of a title to meet its customers needs than do two stores with 1,000 club members each.

This plateau is important to bear in mind, as the home video manufacturers only realize revenue from sales and not rentals. Therefore, although the total rental transaction volume has continued to grow, the prerecorded cassette market, for the software manufacturers, has been a mature one since 1986. In order to show growth in revenue a company needs to either: (1) increase volume; (2) increase prices; or (3) find a new revenue stream. The motion picture companies have utilized each of these options. They have increased volume through a strategy of issuing cassettes at sell-through prices ($14.95-$29.95). However, this can only work with a select few titles. They have also raised prices for nonsell-through titles from $69.95 to $79.95 and then to $89.95. Some titles, such as *Platoon*, are as high as $99.95. Additionally, the software manufacturers are now attempting to create a new revenue stream in the form of commercial advertising on prerecorded cassettes.

In an effort to speed the development of home video as an advertising medium, 12 motion picture companies are working with Nielsen Homevideo Index on a pilot study to measure playback of prerecorded cassettes within the national Nielsen People Meter sample. For this test, the home video suppliers are placing an electronic code into the Vertical Blanking Interval (VBI) of the master tape during the post-production phase. The code will be carried through the duplication process onto every copy of the title released to video stores or sold to consumers. Nielsen has placed special equipment in each household that owns a VCR within the Nielsen People Meter sample. This equipment reads the VBI when the household's VCR is in a playback mode. Each time one of the sample homes plays back an encoded title, the device will identify it and send the information to Nielsen's central computer to be processed with the standard television viewing data. As Nielsen will be utilizing the People Meter, it will be possible to generate exact demographics for prerecorded cassette playback.

There has never been an opportunity to study home video usage behavior in such detail or with this degree of accuracy. At the time of this writing Nielsen had only begun to examine the data. However, it is likely that the availability of this information will change significantly the way the industry views home video and its consumers' behavior.

Conclusion

There is no doubt that the introduction of home video has had major impact on the motion picture and television industries. Nielsen estimates home video revenues from rental and sales in 1987 exceeded six billion dollars. Studio revenue from home video rental now exceeds their box office revenues. More films are being made and a wider range of producers and artists have an increased opportunity to reach a small segmented audience.

However, despite the opportunities for diversity created by the VCR boom, the types of movies that VCR households are renting are not substantially different from those which they previously viewed in the theater. For example, according to the Nielsen VCR report for the fourth quarter of 1987, the top ten rental titles of that quarter were: *Star Trek 4: The Voyage Home*; *Lady and the Tramp*; *Lethal Weapon*; *Crocodile Dundee*; *Blind Date*; *An American Tail*; *Hoosiers*; *Mannequin*; *Project X*; and *Burglar*. How-to and other original products comprised less than 1% of total rentals during the fourth quarter 1987, suggesting that the public has not abandoned print media as its principal source for information. Home video equates with entertainment in the eye of the videophile.

With regard to VCR usage and television exposure, VCR use is small compared to total TV time. Moreover, watching prerecorded cassettes comprised only 23% of total playback activity during fourth quarter 1987. And the videotape rental business is reaching the maturation state of product life cycle. Time spent per VCR household with prerecorded cassettes is likely to remain stable or decline from this point forward.

As with rental, where the video consumer is renting the same product they would have viewed in the theater in the past, VCR owners are recording what they are already accustomed to viewing. Large numbers of households recording a program also equates to large numbers of households viewing it live. Freedom of choice has not opened up a whole new vista, it has only given viewers a greater opportunity to watch what they want to when they want to. What they want to watch is the same as they always have watched.

The 3% of the TV audience that records a program and, therefore, has the potential to zap commercials, will not bankrupt the networks nor is it likely to fundamentally change the way commercial time on television is bought and sold. Recording activity on a per VCR household is on the decline. We may have already seen the VCR have its greatest broadcast impacts.

Indeed, myth has it that because the industry has changed, the consumer must also have changed. But the reality is that the business may have changed, the delivery system may have changed, but to the typical family looking for entertainment, the product remains the same.

4

The Uses and Impact of Home Video in Great Britain

BARRIE GUNTER
and
MALLORY WOBER

In less than ten years the United Kingdom (U.K.) has changed from a country in which only a negligible percentage of its television viewers possessed a home video recorder to a nation in which, today, more than half its television viewers have home video recorders. In family households with children under age 16, resident video penetration (69%) has been even more extensive (Gunter & Svennevig, 1988).

The growth in possession of VCRs happened so quickly that it caught television audience measurement contractors napping. In the U.K., video was having an impact on conventional television viewing and the existing audience measurement technique was not equipped to monitor its usage (Levy & Gunter, 1988). These initial technical problems have been overcome, but the standard audience measurement system is still capable of providing only limited data on home VCR use.

Nevertheless, over the past decade a growing body of research has emerged internationally that has begun to establish a few basic facts about how home video is used. Much of this research, however, has relied upon rather limited self-report methodologies, and these do not always provide a full or accurate picture of VCR use. The broad estimates users themselves are able to supply in self-completion questionnaires or in open-ended evidence during face-to-face interviews may be tarnished by failure of memory or by the wish to project a particular (favorable) social impression. The limitations of such data need always to be borne in mind. Recent evidence has indicated for instance, that continuous diary based accounts of video behaviors do not always match up with self-report estimates (Levy & Gunter, 1988).

In this chapter we intend to address three main issues. In the first part we will attempt to clarify the ways in which we know about the use of VCRs in the home. In order to do this we will draw on the evidence of studies of VCR behavior in the U.K. and highlight contrasts between results obtained through different measurement techniques. Second, we intend to examine the use of VCRs in connection with new television services available in the U.K. We shall look at data on VCR use from households that subscribe to cable television services and data on video recordings that are made of nighttime broadcasts. Finally, we will examine the value of video recordings for viewers. Our principal interest here is whether people with VCRs enjoy their viewing more than people without VCRs and, especially, whether the former attach special value or significance to the programs they record.

PATTERNS OF VCR USE

We will not attempt an exhaustive account of all that is currently known about the character of VCR use. Instead we wish to focus on three questions we believe to be important:

1. Is VCR use characterised by male dominance?
2. Does VCR use promote family togetherness?
3. Is VCR use selective?

The findings presented here are from two sources. One is an annual survey of public opinion toward broadcasting (Gunter & Svennevig, 1988) and the other is from a detailed study of VCR households in four regions of England (Levy & Gunter, 1988). The public opinion survey has been carried out annually for nearly 20 years and includes a core of questions dealing with perceived qualities of television and radio services; but it also contains questions that run for only a few years to examine current issues, one of which is the advent of VCRs. People, therefore, have been asked whether they have a VCR or plan to acquire one and how much they might be willing to pay.

Table 4.1 Who in the Household Generally Buys/Rents the Most Prerecorded Cassettes? (in percentages)

Those who buy/rent	Buy		Rent	
	Men	Women	Men	Women
I do	59	37	60	43
Spouse does	13	27	20	35
Son does	11	14	13	12
Daughter does	6	6	3	4
Other/Don't know	11	17	5	6
	100	100	100	100

The more detailed study of VCR households in which respondents filled out two questionnaires and two, one-week television viewing diaries over two consecutive weeks was conducted among a sample of VCR households in four Independent Television Regions in England—London, Midlands, North-West England, and Yorkshire. The sample was an interlocking quota sample based on sex, six age groups, and four social-groups. From an initial contact sample of 500, usable diaries and questionnaires were received from 446 households giving a response rate of 89%.

Male dominance. Comparisons between men and women in self-reports of video behavior and in diary measures of viewing and video recording behavior revealed some interesting differences. There was only mixed support, however, for the observations of other researchers (Einsiedel & Savage, 1988; Gray, 1986; Morley, 1986) that males tend to dominate in all aspects of home video use.

Based on self-reports, no sufficient differences for example were found between the sexes in the frequency or extent of self-reported purchase or rental of prerecorded video cassettes, with two exceptions. As Table 4.1 shows, there was a difference in the extent to which men and women claimed to be the person in the household who generally buys or rents the most prerecorded cassettes. Men were far more likely than women to say that they personally acquired prerecorded video tapes for viewing at home, either by rent or by purchase. Women were similarly much more likely than men to say that most purchases or rentals of prerecorded tapes were done by their spouses. Both parents agreed that sons were about three times more likely than were daughters to have acquired tapes.

Table 4.2 Numbers of Programs Seen Live, Recorded Off-Air, and Played Back by Men and Women

	Men			Women		
	Week 1	Week 2	2-week Average	Week 1	Week 2	2-week Average
Number seen live	35.1	29.8	32.4	35.8	31.6	33.7
Number self-recorded	1.7	0.9	1.3	2.6	1.7	2.2
Number played back	1.3	0.7	1.0	2.1	1.3	1.7

Although self-reports of video behaviors revealed one pattern of sex differences, diary measures, which reflect a more continuous monitoring of viewing and recording behavior, indicated a different one. Table 4.2 shows how many programs were viewed and recorded off-air and played back by men and women respondents over the two weeks for which diaries were kept. These data show that on average women watched, recorded, and played back slightly more off-air television broadcasts than did men. Thus, on the evidence of continuously self-monitored behavior, women do not appear to lag behind in their personal use of home video.

Still, recording and playback of one's own recordings are clearly not the same as buying or renting and viewing prerecorded cassetttes. So while it is possible that men are more involved than women in bringing in material from outside the home, this does not simply substantiate the hypothesis of a male domination of home video. This is because the present survey indicates that over television output in general, women record and playback more than men do, while other evidence below indicates that rescheduling of this off-air material is more common than the use of brought-in tapes.

Further evidence that it is simplistic to allege an association of video use with male gender, with its implied function of underpinning authority, comes from a study (Wober, 1980) examining the relationships between personality measures, demographic status, and attitudes toward or the actual use of video. This study assessed the degree of welcome for seven technical innovations, including cheap videocassette recorders, and measured the personal attribute of sensation seeking, taking separate account of the principle sense modalities. Fieldwork was by self-completion of special questionnaires attached to viewing appreciation diaries, returned in full by 373 members of a panel in the London region. Answers were weighted by factors calculated to represent a sample of the known population composition, by comparing numbers of replies received—each of 18 cells determined by interlocking two sex, three age, and three socioeconomic status subgroups.

To assess a degree of welcome or unwelcome it was considered unwise to list the innovations without any comment, as likely to yield overestimates of welcome. Instead, for each item (including a new large airport, a satellite with five more television channels, a channel tunnel, more nuclear power stations, pay television meters in every home, and reliable birth control) a possible advantage and possible disadvantage were each cited, the purpose being to prompt a considered rather than a hasty response. The item on cheap videocassette recorders put in parentheses (buy or borrow programs like books or see them when we want; may make cinemas close and harm the quality of television we already have).

On a five-point scale (with 3.0 as a neutral point) VCRs received a positive welcomeness score of 3.4 (reliable birth control led, with 3.9, and nuclear power stations were least welcome, with a score of 2.5). VCR received similar scores from men and women, but much greater welcome form younger adults (4.1) than older ones (2.6) and from those of higher status (3.6) than lower (2.9).

The sensation seeking scales consisted of 16 items exploring liking for stimulating or soothing experiences, and factored with four groups; two factors dealt with venturesomeness in taste and sound, respectively; the other two separated predictability and calm, from preferences for being startled or excited.

Welcomeness for VCRs was particularly linked, in simple correlations analysis, with degree of welcome for a television satellite, birth control, and a channel tunnel, but not so markedly related with welcome for an airport, nuclear power, or pay television. This suggests that desire for technical innovations is not unidimensional, but is likely to arise from other well-entrenched personality structures.

In canonical correlation analysis, two main variates appeared, on the first of which the degree of welcome for videocassette recorders was prominent. This was associated with a welcome for a reliable method of birth control, but not for a television satellite, new airport, nuclear power, channel tunnel, or pay television meters; it was associated with youth, but not with gender; and it was associated with sensation seeking in the realm of foods. The indications from these studies are that the use of the videocassette recorder may just as well, or better, be under-

Table 4.3 Television Viewing and Video Replays in Selected Social Contexts (diary data in percentages)

Watched with.	Off-air Television	Video Playback
Self	24	59
Adults from same household	42	22
Children from same household	9	6
Adults and children from same household	17	6
Visitors	8	7

stood in the context of a perspective of hedonism as in that of a discourse concerned with power.

FAMILY TOGETHERNESS

From what people say about their use of home video, we would be led to believe that video viewing is a social activity that promotes family togetherness. Levy and Gunter (1988) found that 74% of their sample of VCR householders reported that watching a video was an enjoyable way for the family to spend some time together. A smaller proportion (40%) strongly agreed or agreed that "watching video is often an enjoyable way for me and my friends to spend some time together," while a similar proportion (38%) disagreed or strongly disagreed. Diary records revealed in contrast that video use is frequently anything but a gregarious activity. Video householders in the same survey were asked to indicate for each program or video tape viewed whether they watched it alone or in the company of other adults or children either from the same household or who were guests.

The largest slice of off-air viewing (42%) was done in the company of other adults from the same household, while the next most commonly reported viewing context was watching alone (24%), followed by watching with adults and children from the same household (19%). Relatively little off-air viewing (9%) was reportedly done with children only. In contrast, more than half (59%) of all videos played back were watched alone. Viewing video tapes with other adults from the same household was much less commonplace (22% of all replays). Just 6% of replays were with children from the same household (Table 4.3).

Thus, although VCR users readily *claim* that a major attraction of video watching is that it provides an enjoyable way to spend time with family and friends, in practice, this pattern of video use does not predominate. Instead, video playback viewing is most frequently experienced as a solitary activity.

PROGRAM PREFERENCES AMONG VIDEO USERS

VCR users among a national survey of television households in the U.K. (Gunter & Svennevig, 1988) indicated that the programs they were most likely to record were films (68%), followed by soap operas (46%). Other categories of programs were mentioned much less often. Looking at diary based measures of the types of programs recorded and played back in VCR households (Levy & Gunter, 1988) revealed a different picture. Here, movies and soaps were found not to predominate anything like the extent claimed by respondents in questionnaire interviews.

Across two survey weeks, three program types emerged as being viewed most often off-air, namely news (19% of total programs viewed), quiz and game shows (12%), and U.K. drama serials (about 12%). The three program types most often self-recorded were U.S. drama series (15% of total self-recordings), children's programs (12%) and U.K. drama serials and films/TV movies (both 11%). Program types most often played back on video, however, were different again. The three most popular were U.K. drama series (16% of total video playbacks), films/TV movies (14%) and U.K. drama serials (12%).

Summing up so far, we have examined some of the basic characteristics of video behavior drawing upon self-report and diary data. While it is apparent that VCRs are used regularly by most owners, the nature of this behavior varies according to the way in which evidence about it was obtained. Much of what is known about VCR usage derives from self-report data. As we have seen in this chapter, however, the patterns of video behavior indicated by this methodology are not invariably corroborated by evidence derived from continuous diary records of television and video use. The results of this comparison lead us to recommend the use of both methods of VCR usage assessment where resources are available to do so.

Table 4.4 Types of Programs Viewed Off-Air, Recorded, and Played Back (in percentages)

	Seen Off-air	Self-recorded Off-air	Video Playback
U.K. documentaries	7	8	16
U.K. drama serials	12	11	12
U.S. drama series	8	15	8
U.S. drama serials	4	7	10
Films/TV movies	4	11	14
Music and variety	5	8	5
Situation comedy	8	9	9
Quiz and game shows	12	5	4
Chat shows	3	2	4
News	19	4	0
Sport	5	3	4
Documentaries	8	10	8
Serious music/arts	1	0	1
Children's TV	6	11	8
	100	100	100

VCRS AND NEW TV SERVICES

We now turn to our second main focus of this chapter, the place of video in the changing broadcast environment. During 1985 in some areas of the U.K., a number of new television services became available on cable. To date around one million households have been passed by cable but only around 20% of these have subscribed to cable television services.

Although cable uptake so far has not been as extensive as cable operators would like, research undertaken in cable homes has indicated that cable services have a substantial impact on patterns of viewing among subscribers. Comparisons were made between cable and noncable homes in the same areas (AGB Cable and Viewdata/BARB, 1986). Altogether, respondents in the sample watched an average of 4.50 hours of television daily. The average amount of daily viewing was greater in cable homes (5.46 hours) than in noncable homes (4.32 hours). Both the presence of VCRs and of cable television added to the amount of television that people watched. However, VCR use was very slightly greater in noncable homes (0.55 hours daily) than in cable homes (0.51 hours). In cable homes, cable television services jointly garnered the greatest percentage share of total viewing time (40%), more than the amount of time spent with any *single* broadcast channel.

Television viewing in family households with cable and video was slightly more when there were young children around. Cable subscribing households with children (up to nine years of age) and that had VCRs watched 5.58 hours of television daily compared with an average of 5.35 hours of all cable-plus-video homes without children. In general cable channels attained a greater share of total television viewing time in cabled households with children up to nine years of age (50%). The presence of video in cable homes with young children, however, reduced cable's share (40%) with 10% of viewing time being taken up by video playback.

Nighttime Television and VCRs

An extended postmidnight or, in some areas, a full 24-hour broadcasting service became a feature of the television scene in the U.K. in 1988. Some regional ITV companies began showing films, comedies, imported soap operas, and sports in the new nighttime slot. While one ITV company, operating in the London region on weekends, introduced a different style of program, a magazine format designed to appeal to 16- to 34-year-olds.

What does the extension of broadcasting hours mean for viewers' use of television and more particularly their VCRs? Early viewing figures reveal that late-night television in the U.K. still has a long way to go before becoming a consistent and continuing strong feature of broadcasting. The lessons learned from the performance of overnight political broadcasts at election times or of major sporting events indicated that more than just a few insomniacs could be coaxed occasionally into watching television in preference to sleeping. However, attracting loyal viewers on a regular basis is a different matter. Although people may not be prepared to forego sleep to watch television overnight, they may, if the programs are sufficiently appealing, be prepared to record them to watch at a more convenient time.

Judging from audience data produced by AGB for the industry consortium called the Broadcasters Audience Research Board, (BARB), late-night broadcasts seem to have found only *limited* audiences up to 3:00 a.m. However, these results relate only to those people who actually stay up to watch. On weekdays, there are small but significant audiences of between 6% and 2% of adults (aged 16+) up to 2:00 a.m., varying with region. After 2:00 a.m. there is a rapid drop out of viewers. The audiences are noticeably larger on weekends (defined as midnight Friday to 6:00 a.m. Sunday). Between 11% and 2% of the adult population view before 2:00 a.m., again varying with region. These audiences remain up to 3:00 a.m., then drop away sharply. Beyond 4:00 a.m. on both weekdays and weekends the remaining audiences are too small to be accurately recorded.

With a video recorder, it is possible to be a viewer of nighttime broadcasts during the day. Since around half the population possess a video recorder, it is important to find out whether or not late-night broadcasts are being taped. The available data, though limited, indicates that they are. BARB research suggests that the level of recording ITV, although still modest, is of the same order as live viewing after 12:30 a.m. It is unclear, however, whether these recordings are ever actually viewed at a later date. The same information also shows that playback of previously recorded and/or rented video tapes is a serious rival for viewers' time after midnight.

IBA research undertaken in December 1987 used a viewing diary placed with a quota sample of around 1,000 adults (defined by age, sex, and social class) living in the ITV areas then receiving any late-night services, together with a special over-sample of younger adults. Each person filled in a diary for one week showing what media they were using throughout each 24-hour period. This method makes it possible to assess not only television viewing, but also the relationship between television viewing and the use of radio and video. Confirming and adding to the BARB data, this survey reveals sizeable minorities using video, television or radio after midnight and up until 2:00 a.m., then falling away until the morning proper begins at around 6:00 a.m. On an average night, television tends to dominate for most of the time up to 3:00 a.m., but then radio holds its own and has a clear advantage until 6:00 a.m., before the buildup of audiences for breakfast television. One implication of this result is that it is not enough for nighttime television to locate night-people and insomniacs, it must also cause them to break with their established media habits.

In summary, the introduction of extended broadcast television hours and availability on a small scale of new cable television services has had some impact on home video use. Extra channels mean more material to record. In cable homes, however, use of the VCR also means a reduction in direct off-cable viewing. The total amount of time people are prepared to devote to viewing appears to have an upper limit, even though more services are available to view. Over-night television has made inroads into audience viewing habits. Only small audiences bother to stay up very late to watch television—though rather more do so on weekends when they do not have to get up for work the next day. Even so, the most common habit seems to be to record late night programming to watch at a more normal viewing hour. Thus, the VCR gives viewers control over the ever increasing volume of television output. Through VCR they can continue to watch when it is most convenient for them to do so.

THE VALUE OF VIDEO

In this final section of our chapter, we want to examine how people feel about video, how that affects their behavior, and the implications of those questions for certain aspects of broadcast policy. Home video is undoubtedly a valued commodity. Users realise how much more control it affords them over what they can watch and when. Once acquired, the VCR tends to be used regularly by nearly all its users. VCR users generally view off-air recordings fairly quickly so that tapes can be recycled and used for future recordings. Some VCR households have been found to build their own tape libraries, a practice which is, strictly speaking, illegal. The significance of video for most people thus seems to be that it gives them greater control over when to watch television programs and films. The most serious problem people report having with their home videos is lacking the time to view all the programs they tape (Levy & Gunter, 1988).

Aside from these findings, however, to what extent do people value their recordings compared with other programs? Are the programs people choose to record those which mean the most to them? To what extent does the use of home video depend on the surplus of valued programs broadcast at the same time on the available television services? Recent additional analyses of video household data collected for the IBA have shed some light on these questions (Wober, 1988).

The way in which the foregoing and subsequent studies were done is one which had been established for over a decade. The procedure is to recruit respondents by interview, who are then sent a diary and associated questionnaires. The diary lists every broadcast program, and asks for a mark, on a six-point scale, of appreciation for each and every program of which enough has actually been seen in order to form a judgment. In this way programs are rated on a self-assessed criterion of sufficient experience in which to record a judgment; it is possible to count the number of programs that were "seen" in this way, which may mean that they were seen in their entirety or less. Responses from the sample audience for each program are transformed by calculation to a position on a 100-point scale, in which 100 denotes extremely interesting and/or enjoyable and 0 corresponds to not at all interesting or enjoyable. Variations in this procedure have been used, exploring the point scales labeled only at their top and bottom extremes, and separating enjoyment from strength of impression made upon the viewer. The procedure thus gives measures of behavior, in terms of number of programs viewed, which can be counted in each of several different genres, and of what this experience has felt like.

It has long been realised that diary respondents will readily reply to additional questionnaires about broadcasting—and other—issues; and the replies to these questionnaires can be related to the experiences and use of broadcasting by simply correlating them with data taken from the diaries. An example of this procedure has been described, above, in the study on degree of welcome for VCR amongst other technical innovations.

A two-week diary was set up with which to examine patterns of use of VCR (Levy & Gunter, 1988). The monograph presenting the results of this work is preoccupied with behavioral details and the evaluative information has remained available for further analysis. Some of this analysis is presented below. For each program viewed, a score is available on the strength of impression it made upon the viewer; these scores can be expressed on a scale from 0-100 and calculated for whatever was seen (live, not on VCR) within each of thirteen program types.

The first column in the table shows the average score on the strength of impression scale, for each of the program types; these reported experiences all pertain to what was viewed at the time it was broadcast, on air. The documentary and magazine program type included 14% of their number which were recorded onto cassette; these items within this type which were recorded, had impression scores which were on average ten points greater than the average for the category as a whole. It so happens that only 14% of the documentaries that were seen were recorded; the impression score for the documentaries that were recorded stood at 77, while for the majority (86% of this category) that were not recorded the average impression score was 65.

It can be seen, therefore, that for all categories where any of what was seen was recorded, the recorded items are ones that were given higher impression scores. This demonstrates that people tend to record material that has made a greater impression on them. This study did not obtain impression scores at the point of replay, so whether or not viewing at replay earns higher impression scores than average for item viewed on air, in the same type or genre, remains to be confirmed. It is clear that some types have a higher average impression score than do other types; but from the percentages that are recorded from each type there is no reason to think that a greater proportion is recorded from the best appreciated type than is saved from the least appreciated type. What is clear is that those items from within each type that people record, whether from best liked or least liked program types, are the most impressive items of their respective genres.

This particular study did not include people who did not own VCRs; so it is not possible to use it to shed light directly on such matters as whether people with VCRs will (other things being equal) enjoy their viewing more than will those without VCRs. Nevertheless, it seems to indicate that the VCR is being used, at least as a "short-term memory" to keep material that is more striking, perhaps to see it again or for other members of the household to see, who may not have been at the first viewing.

If having a VCR is a way toward a better planned system of viewing, then a combination of using the recorder and having more channels to choose from (which are being prepared for initiation in Britain in 1989; some plans mention seven new English language channels, others speak of a dozen, carried on two separate satellite systems) should further increase the overall satisfaction with the viewing diet. One may also expect that people who have no VCR would correspondingly lose out on opportunities to see items that clash with what they say they want to watch.

In another survey (Wober, 1988) questions were placed on BARB's national television opinion panel. A sample of 2,524 people answered questions on the "losses of choice." The aim here was to find out to what extent people felt that clashes occurred between two or more programs they really wanted to watch. And then subsequently, whether the making of video recordings was connected with the perception of a surplus of desired program material. For 20 half-hour time slots on one

Table 4.5 Impact of Recorded Programs

Type	Seen Live I Score	% That Have Been Recorded	I Score for Recorded Items Over All Items Seen
Documentaries, magazines	67.1	14.0	10.3
Sport	64.2	9.6	7.8
Art	63.2	—	—
U.K.-made series	62.7	12.3	7.0
News, current affairs	62.5	5.0	12.3
Non-U.K.-made serials	62.4	17.8	12.5
Variety, comedy	61.4	13.8	10.2
Situation comedies	60.3	14.8	14.1
U.K.-made serials	59.9	18.9	13.8
Films, TV movies	58.9	14.4	9.4
Game Shows	58.9	8.3	6.6
Non-U.K.-made series	56.8	17.6	8.0
Children's programs	56.7	23.6	12.4
Other light entertainment	56.1	3.8	2.5

day of the week respondents were asked to choose one out of five responses. Taking all four channels into account, did they think that:

5 = there were 3 or 4 programs on at the same time, that I wanted to see;
4 = there were 2 programs on at the same time, that I wanted to see;
3 = there was 1 program on that I wanted to see, and that I did see;
2 = there were NO programs on that I wanted to see, but I did watch all the same;
1 = I was not at home or at home, but not watching.

Respondents were presented with a list of time slots, in half an hour units from 3:00 to 12:00 p.m. and asked to say for each slot, what the contents were in terms of the above question. Further questions dealt with VCR availability and use.

Table 4.6 shows the results of the contents of time slots, condensed to show major time bands *within* each of which there was a relative similarity as contrasted with the differences between these time bands. Before 5:00 p.m. nearly four out of five people indicate that they are not at home, or, if at home, not watching television. This is the basis on

Table 4.6 Perceived Range of Choice Among Available Viewers (in percentages)

Time band	Those who said choice was:			
	Oversufficient	Sufficient	Insufficient	
3:00 - 5:00 P.M.	13.6	63.1	23.3	100.0
5:00 - 11:00 P.M.	17.7	64.3	18.0	100.0
11:00 - 12:00 P.M.	23.3	53.5	23.3	100.0

which the data on "over- and insufficiency" have been calculated, to focus on times when people were available to view.

The major and striking point to be noted from these figures is that there are *as many people* who say there was "oversufficient" choice (that is to say, two or more items on *at the same time*, both or all of which they would have liked to have seen) as there are those who say there was insufficient choice. This shows that when people are involved in viewing, from half to two-thirds of the replies indicate that—in the present four channel system—there is sufficient choice. Before the main evening's viewing there is some predominance of insufficient above the level of oversufficient choice; but during the peak time band and the hour afterwards, there is just as often an oversufficient as an insufficient choice.

What can be done with oversufficient, or surplus, choice? One solution for the viewer is to record some of this. Over three in five people have the use of a VCR; over two in five people say they record something at least three times a week; and over half those who have the use of a VCR say they play back everything they have recorded. Another quarter of the VCR owners claims to play back four out of every five items recorded. Women are not much different from men in their reported availability and use of VCRs; younger people seem more likely to have a VCR, but less likely to use it for recording something than are older people; while upper status people are more likely than those of lower status to have, and to use, VCRs for recording.

The next questions to examine are whether people who believe there is a surplus of available material are more likely to be equipped to make use of this material and whether they are more likely actually to *use* desirable material.

To examine these questions, three scales were created. The first scale refers to the time band 3:00-5:00 p.m. Anyone who said there were two programs on at the same time both of which they wanted to

Table 4.7 Extent of Surplus Availability (in percentages)

	Time Bands		
	A 3-5 P.M.	B 5-11 P.M.	C 11 P.M.-12 A.M.
Base: All, including regional booster samples:	3,284	3,284	3,284
Reports *no* surplus at all	96	70	—
At least 1] point	2	11	2
" " 2] on	2	7	1
" " 3] the	—	4	—
" " 4] surplus	—	2	—
" " 5] scale	—	2	—
" " 6 or over] (see p. 61)	—	4	—

see was given a score of 1. Those who said there were three or four items they wanted to see that were broadcast simultaneously were given a score of 2. Thus for the mid-afternoon time band a person could score 0 (for no surplus availability reported) up to 8 (all four slots have two or more "surplus" items). Similarly, a scale was set up for the broad prime-time band of 5:00-11:00 p.m., with scores ranging from a maximum of 24 to a minimum of 0, and likewise for the late evening from 11:00 p.m. to 12:00 a.m., with a maximum of 4 and a minimum of 0.

The main point to observe here is that almost one in three (30%) report there was surplus material they would have liked to have seen, in the 5:00 - 11:00 p.m. time band. In Table 4.6 it was shown that 18% of available viewers report surplus, but that was an *average* result, per half-hour slot. What is shown here is something that is *cumulative*—which accumulates anybody for whom this experience was true, at any time across the whole time band. There are more than one in ten of the population who indicate, also, that there are three or more half-hour slots among the twelve across the time band, for which they judge that there is surplus choice.

We know that certain numbers of people say they record and play back programs different numbers of times a week. But the next step has been to give each person a score for the amount of *recording* they do (score 0 for not having a VCR, score 1 for having one and using it hardly ever, up to score 5 for saying they record seven or more times a week. Likewise, a *playback* scale was set up, and finally, these indices

Table 4.8 Correlations Between Reports of Surplus and Steps Taken to Use It

		Surplus Reported in Time Bands			VCR Behavior		On-air Weight of Viewing
		A	B	C	Record	Playback	
Surplus scores	A	1.0	.51	.24	.11	.02	.06
for time bands	B		1.0	.38	.20	.02	.09
	C			1.0	.07	.07	.07
Record					1.0	.15	.17
Playback						1.0	.06
Numbers of respondents involved in the correlations:		3,284	3,284	3,284	2,020	1,987	1,987

were correlated together, with measures of reported surplus and a three-point weight of viewing index (0-17 hours a week taken as light viewing, scored 1; 18-31 hours taken as medium viewing, scored 2; and 32 hours and over, scored 3 for heavy viewing).

Several points emerge from Table 4.8:

- Anybody who reports a surplus in one time band, also tends to report surplus in other time bands.
- Those who tend to report a surplus, also say they record material more often (the coefficients under the column Record are all highly significant, that is, not equivalent to zero).
- However, those who tend to report a surplus do NOT also tend to claim that they play back what they have recorded more often than those who are less inclined to talk of surplus.
- Heavy viewers are more likely than light viewers to report a surplus. This relationship is not very marked, but it exists. Heavy viewers are more likely to say they record and even play back more material than are light viewers; but the relationship with recording is clearly greater than that with playback.

In the foregoing discussion, a preliminary question has been posed and investigated. With four channels already available, and regulated program schedules designed to reduce program type clashes and to increase complementarity and choice, how much is there already of a *surplus*, and how much is there of *dearth*? With one channel it must be expected that dearth—not having something one wants to see, when one wants to view—will be quite common. Surplus is not possible. With two channels, surplus becomes possible, and dearth should be reduced. This should be even more evident with four channels.

In practice, it is clear that for the main viewing time band there is just as often a surplus of programming as there is a dearth. If, then, one has a VCR, one may record some or all of the surplus; and, of course, one may play it back. Many people do have a VCR, and many report its frequent use. What is more, the people who say they do more recording are indeed the ones who report more surplus. These people are screen hungry as well as well-screen-fed, since heavy viewers are not only greater program consumers, but also tend to report more surplus than do lighter viewers.

What is noteworthy, however, is that there is *not* a significant relationship between amount of surplus reported, and amount of VCR playback. This means that people who say there is greater surplus (they are heavier viewers and would like to see more programs than they can watch at any one time) do not tend to use the material in playback any more than do those who say there is little or no surplus. In short, there is waste at two levels. At one level there are more programs that avid viewers want to see than they can see (and which they try to conserve, by recording them); but at the next level, having recorded more programs, these avid viewers are not so likely to play them back. It must be suggested that one reason for this is that there is a continuing supply of good programs, which people continue to want to see. Some people probably record over what was previously recorded but not seen. One of two limitations must be that a ceiling is set to the time for which people are willing and able to watch the screen. The second limitation is that as people become more selective in assembling and viewing what they intend to be noteworthy experiences, some of these programs have to be cast in the role of background, against which the more prominent items stand out as "figures."

It does not necessarily follow that more channels will bring more choice or that the use of VCRs will establish a better planned and more satisfying—and perhaps healthier—experience of viewing the screen. Economic forces tend to mold competition in terms of like materials competing against like. Left to themselves, if one or two channels are finding large audiences at the same time with soap operas, thrillers, or situation comedy, a third or further channels will tend to compete by offering similar products, rather than by showing something different such as documentary or arts programs. Only in a socially regulated system is schedule complementarity pressed upon the providers. So, if more channels without such controls become available, viewers will experience one kind of waste—that is, they will miss seeing programs of a similar type to what they are watching. If more channels become available together with some mechanism for promoting variety by program type, at any given time, viewers will experience another kind of waste. Here, they will miss seeing programs of different types. In short, reducing dearth is accomplished at the cost of increasing surplus, or waste; and the only observable outcome in terms of satisfaction or appreciation that can be measured with the same instrument before and after a change suggests that people *adapt* to the array of what is available to such a degree that they evince no greater satisfaction with greater than with lesser program availability.

In a decade of rapid, unprecedented change in the home media environment, perhaps the most significant development of all, especially where the average television viewer is concerned, has been the growth of home video. It is now unusual *not* to have a VCR in Britain. With the recent expansion of available television services, and much more promised by the early 1990s, the establishment of home video has given viewers the power to control what and when they watch programs that appeal most even in this increasingly competitive television environment.

Viewing can be done around the clock in most of Britain. This is a very recent development. It means that viewers could be drawn to watch programs at times they would not normally have considered for viewing. The alternative is to miss programs. With VCRs at their disposal, however, they can capture programs shown at awkward times and hold onto them until they are ready to watch.

More television does not seem to mean more viewing. Many people appear to have a fixed time budget to invest in watching television. Even with the spread of television sets throughout the household, making it possible to watch in whichever room in the house one happens to be, the total number of hours devoted to paying attention to the small screen is allowed to stretch only so far. With more material to choose from and fixed viewing time, coupled with largely unvarying viewing periods during the day, the ability to capture programs from any channel at any time may become increasingly important. Rather than being drawn by television as it spreads to cover more channels and more time, viewers through their VCRs will be able to draw from it, thus maintaining an important balance where control over viewing behavior really lies.

REFERENCES

AGB Cable and Viewdata (1986). *The 1986 cable monitor*. London: AGB Cable and Viewdata Ltd.

Einsiedel, E. F., & Savage, D. (1988). *VCR usage patterns among a rental segment: Expanding definitions of technology use in a Canadian sample*. Paper presented to the International Communications Association meeting, New Orleans, Louisiana, May 29-June 2.

Gray, A. (1986) *Video recorders in the house: Women's work and boys' toys*. Paper presented to the International Television Studies Conference, London.

Gunter, B., & Svennevig, M. (1988). *Attitudes to broadcasting over the years*. London: IBA and John Libbey.

Levy, M. and Gunter, B. (1988). *Home video and the changing nature of the television audience*. London: IBA and John Libbey.

Morley, D. (1986). *Family television: Cultural power and domestic leisure*. London: Comedia Publishing.

Wober, M. (1981). *Pyramids or chariots—The satellite question*. London: Independent Broadcasting Authority, Special Report.

Wober, M. (1988). *The cost of choice—A calculus of programme want, variety and waste*. London: Independent Broadcasting Authority, Research Paper.

Part II
Using the VCR

5

Adolescents and the VCR Boom: Old, New, and Nonusers

BRADLEY S. GREENBERG
and
CAROLYN LIN

The rapid diffusion of cable television and videocassette recorders in the 1980s marks a new era in the entertainment media scene. The abundance of television content available by means other than standard commercial broadcast television greatly enhances viewer choices. The slick technology accompanying these television developments provides new and greater freedom over what, when, and how to attend to television. And smack in the middle of this is a generation of young people for whom these technologies, with their accompanying bells and whistles, are increasingly commonplace and normative.

More than half the homes in the U.S. now have VCRs, more than half have cable and half of those have added a pay channel (Television Information Bureau, 1988), whereas in 1985, Greenberg and Heeter (1987) reported VCR penetration in homes including adolescents at 33% in a Michigan urban site. A recent suburban Boston study (Hughes & Dobrow, 1988) found that homes with children are even more likely to have VCRs (85%) and to have cable (60%). This video culture emphasis has found favor among adults as well; in an adult survey in Lansing, Michigan, Lin (1988) found that two-thirds preferred watching videos to going out to the movies, and more than half had joined a video club. In addition, more than half reported spending more time watching television and watching it more with their family after obtaining a VCR.

The audience measurement from AGB stated that U.S. children under 18 average 3.4 hours per week watching VCRs, a rate double that of adults, that those same youngsters spent 61% more time viewing prerecorded tapes than did adults, and that they were considerably more likely to be watching on a VCR during primetime than watching a pay cable program (Teitelbaum, 1988). Hughes and Dobrow (1988) indicated more than twice that level of VCR use; their Boston teenagers averaged 1.2 hours per day watching tapes and 1.5 hours making new ones. In terms of content preferences, their sample most preferred rental movies. The youth sample claimed that they went out to the movies less often, read less, and watched broadcast television less. Greenberg and Heeter (1987) found more VCRs in up-scale homes with Michigan teenagers—the parents had more income and education—there were more television sets, more basic and pay cable television and more personal computers, as well as more television sets claimed as "mine" by the teens. This sample watched television series with more sexual content, went to more R-rated movies, and watched more R-rated movies on their VCRs. They watched more soaps and more primetime shows, although a summary measure of daily viewing did not yield different total time estimates.

This is not only an American phenomenon. Roe's study on Swedish adolescents (Roe, 1987) identified 7.5 hours per week of VCR use in VCR homes and 2.5 hours among teens without VCRs at home. He also indicated that more frequent VCR users came from lower SES backgrounds and had lower school achievement. Roe also reported strong content preferences among Swedish youth for explicit violence, horror, and sex.

Teenagers receive little parental guidance with their normal television activity (Medrich et al., 1985), and this does not change when a VCR is added to the facilities. The increase in R-rated movie experiences via VCRs was found among adolescents who reported little parental intervention, monitoring, or guidance (Greenberg & Heeter, 1987), a not surprising linkage. And as the number of single-parent households and homes with two working parents soars, the latch-key children watched more television (Brown, Bauman, Lentz, & Koch, 1987); Lin and Atkin (1988) found that the number of employed parents as well as single-parent households were linked to more limited parental oversight of media activity. Brown and her colleagues contended further that access to cable television and to VCRs is predictive of time spent viewing, but not of parental mediation. There is support for that contention; parental mediation of children's viewing was less prominent in basic cable homes (Haefner, Hunter, & Wartella, 1986) and in pay cable homes (Atkin, Heeter, & Baldwin, 1987).

Alongside these changing viewing patterns and viewing contexts with VCRs, and with the knowledge that parents are perceived as doing very little intervention, the issue of viewer motives comes into play. Roe (1983a and 1983b) showed that Swedish adolescents were using VCRs to reinforce peer group autonomy, to circumvent censors, and to rebel against adults. Rubin and Bantz (1987) proposed that different age and gender groups use VCRs for a variety of purposes, of which two major ones are time-shifting and the ability to control more aspects of viewing. They also suggested that the motives, functions and/or gratifications of VCR use correspond to television use. But, of course, the VCR is television, clothed in a different garment and perhaps capable of more changes of costume, depending on what the video clothing store has available that day or what the user has stashed in the video closet at home. The question may not be whether the gratifications sought and obtained are different, but whether they are capable of being more fully satisfied within the VCR context.

There also has been much interest in just how active the VCR user is or becomes. Levy has been a prime examiner of audience activity (with adults) and has tracked activity throughout the media use process. Levy (1987) found VCR users to be "selective, somewhat involving and often useful" in the pre-, during-, and post-exposure periods. He suggested that the nature of the media technology linked with program content may be associated with different degrees of audience activity. Similarly, Lin (1987) found that adolescents who were more active in cognitive, affective, and behavioral ways with new media were more gratified by their media experiences.

Another aspect of audience activity takes the form of the control that VCRs provide over viewing activities, conditions, and program options (Levy & Fink, 1984). But the exercise of control requires involvement and activity from the user—planning for time-shifting, zapping commercials while taping or zipping through them while viewing, making selections for taping, and programming the machine. The little evidence available indicates that adolescent users may be doing some of these with a vengeance—Hughes and Dobrow (1988) found that 96% of their teen sample turned down the volume and zipped through commercials during the playback of tapes. Lin (1988) reported that teenagers who have more control over what they can rent or tape and

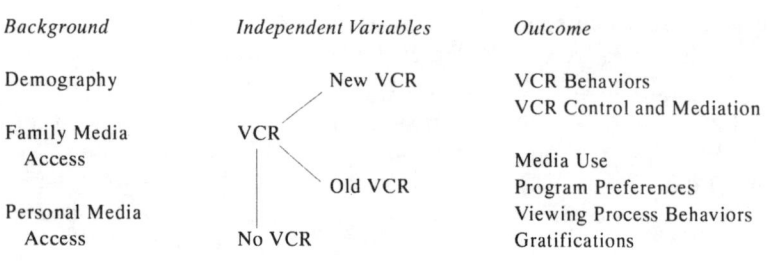

Figure 5.1. Study Components

when they can do so are more active viewers. Thus, greater control and activity in the access phase is collated with greater activity in the viewing phase.

Research on new media is often out of date by the time it is designed, let alone reported. In studies done between 1985 and 1988, VCR penetration doubled; it is increasingly difficult to find non-VCR adolescents in sufficient, representative clusters. On the other hand, that means that some adolescents have had VCRs available to them for several years, while others are relative newcomers. Thus, the research questions can be extended from comparisons of those with and without VCRs to those of the same age who had video for different lengths of time.

Figure 5.1 graphically displays the key issues discussed to this point. It identifies the strategy to be used in making comparisons among adolescents with and without VCRs and can be used as a guide through the remainder of this chapter.

This study will differentiate VCR and non-VCR adolescents and then distinguish between recent and less recent adopters.

Demographic distinctions will be examined primarily in terms of family structure characteristics and socio-economic status. Then, the subgroups will be considered in terms of other media they have available.

Media use behaviors, program preferences, activities while viewing, parental mediation, and gratifications sought and obtained provide a general set of outcome variables for all the study groups. For those who have VCRs, their patterns of VCR use, zapping or zipping of commercials, and the nature of parental control and mediation will be analyzed.

METHODS

In-school questionnaires were administered to 206 seventh graders and 221 tenth graders in Haslett and Holt, Michigan, two suburbs of Lansing, in spring, 1987, after pretesting with 43 high school students in a nearby school district. In this chapter, the data from the two age groups are combined.

Given this chapter's specific interest in comparing VCR and non-VCR youth and its extended focus in comparing youth whose families recently acquired VCRs with long-time VCR families, identifying the VCR penetration in this youth group is important. VCR penetration was 78%; new or recent owners of two years or less duration were 59% of the VCR subgroup. Thus, for most comparisons, the analyses will be comparing more than 330 VCR youth with approximately 100 non-VCR youth, and 200 recent or new VCR acquirers with 130 more established VCR homes.

Demographic attributes. The focus here was on gender, family composition, and socioeconomic status (SES). Family composition was assessed with the size of the family, the number of siblings and the parental structure, i.e., an original mother and father, mother only, mother plus stepfather, etc. SES was assessed by determining the numbers of cars and bedrooms at the respondent's address, whether the residence was owned or rented, and the whether the respondent had his or her own bedroom. In addition, the respondents reported their average school grades (A to F).

Access to other media. Interest in media available in VCR and non-VCR households was approached in two ways—those media available to all household members and those media "owned" by the respondent. For *family access,* questions determined the number of television sets and color television sets, and whether the household contained basic cable television, (and pay channels if there was cable), a push button television set, a remote control television, video games player, video camera, videodisc player, compact disc player, and a personal computer. For *personal access*, the items for the youngsters included whether each had his or her own television set, audiocassette player, stereo system, record player, Walkman, telephone, and calculator.

Use of media. For nontelevision media, respondents estimated the daily minutes they read newspapers and magazines and the daily hours they listened to the radio and to music on nonradio outlets. For television, they gave separate estimates for the amount of time watching television with their family and alone, including independent estimates of such viewing on schooldays, Saturdays, and Sundays.

TV programming preferences. Television program types examined were soap operas, detective shows, situation comedies, dramatic series (e.g., "LA Law"), movies or miniseries, sports, music video shows, advice column shows (e.g., Dr. Ruth), and news and public affairs programs. A five-point scale (Very Often, Often, Sometimes, Rarely, Never) was used to assess how often they usually watch each program type. For respondents with cable television, additional questions asked how often they watched unrated, G, PG-13, PG, and R-rated movies, with a brief definition of each rating category provided.

General television viewing process behaviors. This segment examined certain behaviors of the young viewer that occur before, during, and after watching television. It also determined what kinds of mediation of television viewing occurred in the household and the nature of the youngster's participation in the viewing selection process. All scales except where noted were the same five-point, "very often/never" type. Scale construction of all indexes used is described more fully in Lin (1987).

Specifically, two scales of television pre-exposure behavior were used. The first, "planning," had three items to assess the extent of advance planning in television selection, e.g., "I know ahead of time.... what television show I want to watch." The second, "orientation," included three items to determine how much of a habit television viewing was, e.g., "Watching television is part of my daily activities," with responses on a five-point, "strongly agree/strongly disagree" scale.

During-exposure behaviors were gauged with four scales. The first scale of four items assessed their emotional "involvement" while viewing, e.g., "I get into a show that I am watching." Three other scales focused on different aspects of their technical involvement while viewing: (1) "zapping" of commercials, e.g., "When commercials come on during a show, I change channels until they are over"; (2) "channel switching," e.g., "During a show, I switch channels to check what else is on"; and (3) "multichannel viewing," e.g., "During a show, I switch channels to watch two or more shows at the same time."

Two scales assessed postexposure activity. The first, "involvement," used four items, including, "After watching an interesting show, I think about it for a long time." The second, "media motivated activities" with four items, asked questions of this order, "I will go do something that is fun for me because I saw it on television."

Parental mediation was measured with regard to controls exercised by parents over specific program choices and as implemented by more general family rules. A "parents' mediation" scale used three items, such as "My parents would encourage me to watch certain television shows." A "family rules" three item scale included, "In our family, we have rules on how late I can watch television." The extent to which the child could manipulate the viewing environment (i.e., viewing flexibility) was examined in terms of his/her involvement in program "decision making" and "technical control" while viewing. The former used four items including, "I decide what shows to watch while watching with my family," and the latter had six items, including, "When I am watching with my family, I can control whether to change channels during a show to find a more interesting show."

Respondents also indicated in different portions of the questionnaire the gratifications they were seeking ("I watch television to...") and the extent to which those same gratifications were obtained ("I am satisfied with..."). Six gratification factors were examined, each consisting of three or four items summed together: (1) Surveillance, e.g., "...to find out about the latest news on popular music"; (2) Informational guidance, e.g., "...to get advice on how to solve my personal problems"; (3) Entertainment, e.g., "...want to have some fun"; (4) Diversion/escape, e.g., "...to forget about my problems"; (5) Interpersonal communication, e.g., "...to find something interesting to talk to my friends about"; and (6) Parasocial identity, e.g., "I like to think of some people on television as friends."

VCR viewing process behaviors. Among those with VCRs, some behaviors especially related to VCR viewing were also investigated. First, *prior planning* asked in three items whether the youngsters decided ahead of time "what shows to tape while I am watching television." Second, *zapping* with VCRs was described with two items, "When I am taping a show, I tape the commercials too," and "When I am playing back a show taped earlier, I fast forward to skip commercials."

A third set dealt with the viewing of prerecorded and home-made tapes. Respondents provided estimates in hours of watching (1) videos taped at home from Monday through Friday; (2) videos taped at home Saturday and Sunday; (3) videos taped at friends' homes in a typical week; (4) videos rented and watched at home in a week; and (5) videos rented and watched at the homes of friends in a week. A final set of questions asked what program types they taped, using the ten categories identified earlier in terms of program type preferences.

Mediational and decision-making opportunities in the general television viewing situation have their parallels in the VCR viewing context, and these also were re-examined for VCR owners. "Parent mediation" used four items, including, "My parent(s) would tell me not to watch certain types of videos," and the "family rules" index was four items, including, "In our family, we have rules on what kinds of videos I can rent."

Decision making extended across four factors—what kinds of videos to get, how many to get, when to play them, and whether to zip the commercials. *What* consisted of such items as, "I can decide what videos to rent from a video store for myself"; *how many* included, "I can decide how many videos to rent for myself"; *when* included, "I can decide when to play back videos for myself at home"; and *zipping* included, "while my family is playing back a video, I can decide whether to fast forward the commercials."

FINDINGS

These data have been organized to make two comparisons: the first is between youngsters with and without VCRs in their homes and the second is between those who have had VCRs for two years or less (identified as *New VCR* in the tables) and those who have had VCRs for more than two years (*Old VCR*). This split maximized the number of respondents in each group, and additional categories would have produced skewed distributions.

Demographic attributes. VCR adolescents have more brothers and sisters and are more likely to live in traditional families with an original mother and father; the overall household tends to be larger (p<.08). Further, homes with adolescent boys are more likely to acquire VCRs than homes with adolescent girls. Fully one-third of the non-VCR adolescents have but one parent in their household, in contrast to 15% of the VCR homes. It follows then that there are clear SES differences, using indirect indicators; those with VCRs have more cars in the family, more bedrooms in the house, and are more likely to be in owned, rather than rented living quarters. There are no differences in self-reports of school grades.

There are fewer demographic differences between the new and the old VCR users. However, those differences found—more of the old

Table 5.1 Demography by VCR Access

Demography	No VCR (n=87)	VCR (n=332)	New VCR (n=198)	Old VCR (n=133)
a. Size of family	4.1	4.4	4.4	4.5
b. Number of siblings	1.2	1.4*	1.4	1.4
c. Parent structure				
Mother and father	53%	63%*	58%	70%*
Mother only	33	15	19	11
Mother, stepfather	6	12	11	14
d. SES				
# of cars	2.2	2.6*	2.5	2.8*
# of bedrooms	3.3	3.7*	3.6	3.7
Own their home	69%	87%*	85%	89%
Have own room	83%	88%	85%	89%
e. Gender: % female	59%	45%*	48%	43%
f. Grades in school	7.2	7.3	7.4	7.2

*The asterisked comparisons indicate statistically significant differences at $p < .05$, using two-tailed tests.

VCR users have intact parents and there are more cars in those families—are consistent with the VCR/non-VCR comparisons.

Media access. The family that has acquired a VCR has added it to a larger collection of media capabilities in that same family, as Table 5.2 shows. The VCR adolescent has more television sets, more color sets, video game machines, compact disc players, and video cameras. Similar differences also are found in the comparison between new and old VCR youngsters. Old VCR respondents have more hardware than recent adopters, e.g., video disc players, personal computers, and push button television sets.

Of note also is the absolute level of media ownership claimed among these homes. In homes averaging nearly three television sets with two of them in color, a majority now have video game players, personal computers, cable television, and television sets with push buttons and remote controls. One-fourth claim video cameras and one-fifth compact disc players.

Personal media, to which the youngsters claimed to have their own, private access, discriminated VCR households in terms of a television

Table 5.2 Family Media Access by VCR Access

Family Access	No VCR	VCR	New VCR	Old VCR
a. # of TV sets	2.3	3.1*	2.8	3.5*
b. # of color sets	1.5	2.2*	1.9	2.6*
c. Video game player**	61%	75%*	71%	79%
d. Video disc player	9	15	11	22*
e. Compact disc	10	20*	16	25*
f. Video camera	14	24*	21	28
g. Personal computer	51	53	48	62*
h. Basic cable TV	69	75	73	78
Cable movies	76	77	80	73
i. Push button TV	56	63	58	70*
j. Remote control TV	56	63	58	71*

*The asterisked comparisons indicate statistically significant differences at $p < .05$, using two-tailed tests.
**(% YES for items c-j)

set, an audiocassette player, and personal calculator; only the television set difference was maintained between more recent and more established VCR homes, as in Table 5.3. But for all the technologies inquired about—including stereo systems, record players, Walkmans and telephones—more than one-half of all adolescents surveyed said they had their own.

Media use. Mass media use, both print and electronic, was not different between VCR and non-VCR adolescents, nor was it different between Old and New VCR adolescents. Further, no differences were found in estimated newspaper or magazine reading time, in radio listening time, record playing time, or in television time—whether in family or private viewing situations, or on weekdays or weekends. Clearly, the electronic media dominate media time; print reading consists of a half hour or less on a daily basis, compared with four hours for television viewing on schooldays (4.8 in old VCR homes), 2.5 hours with radio, and nearly two hours listening to records.

TV program preferences. Having or not having a VCR does not influence preferences for different types of television programs. Among ten different program types, only one difference was noted—a greater preference in VCR homes for police/detective series, but such a difference can best be attributed to chance, given the large number of comparisons. Tendencies were found of stronger preferences for movies ($p < .06$) and music programs ($p < .08$) among those in the older

Table 5.3 Personal Media Access by VCR Access

Personal Access (% YES)	No VCR	VCR	New VCR	Old VCR
a. TV set	46%	59%*	52%	70%*
b. Audiocassette	59	73*	70	77
c. Stereo	63	68	66	70
d. Record player	61	68	66	70
e. Walkman	79	84	83	87
f. Telephone	49	52	53	50
g. Calculator	77	87*	84	91

*The asterisked comparisons indicate statistically significant differences at $p < .05$, using two-tailed tests.

VCR homes. All adolescents most strongly preferred situation comedies, followed by movies, sports and music shows; advice programs were least preferred.

More meaningful is the difference obtained within the subset of homes with cable television movie channels. There, the preference for R-rated films in VCR homes was significantly greater than in non-VCR homes, and this difference persisted with old VCR homes having an even stronger preference for R-rated films than new VCR homes. Furthermore, among movie types, unrated and G-rated movies were least preferred.

Viewing process behaviors. Differences in viewing process behaviors are minimal. Pre-exposure behavior comparisons showed that the orientation to television watching as a habit was strongest among the more established VCR adolescents, and there were no differences in the planning of television viewing. Planned viewing, overall, was reported as a quite frequent activity (3.6 on a 5-point scale). During exposure, VCR youth claimed to zap commercials more regularly, and those in older VCR homes tended to do more channel switching during programs ($p < .06$), while multichannel viewing was not different. Post-exposure behaviors showed no difference in involvement nor in media-motivated activities.

Parental mediation of particular program choices was the same across all these youth groups; however, there were more general family rules about watching television in non-VCR homes than in VCR ones. Overall, mediation of any kind was the least frequently reported viewing process behavior. In contrast, the adolescents said that making their own viewing decisions was their most frequent viewing process

Table 5.4 Viewing Process Behaviors by VCR Access

Viewing process behaviors[a]	No VCR	VCR	New VCR	Old VCR
1. Pre-exposure				
a. Orientation	3.1	3.2	3.1	3.3*
b. Planning	3.6	3.6	3.6	3.6
2. During exposure				
a. Involvement	3.2	3.1	3.1	3.2
b. Zapping	3.0	3.3*	3.2	3.4
c. Channel switching	3.1	3.0	2.9	3.1
d. Multichannel viewing	2.3	2.3	2.3	2.4
3. Postexposure				
a. Involvement	3.0	2.9	2.9	2.9
b. Media-motivated activities	2.3	2.4	2.4	2.5
4. Mediation				
a. From parents	2.4	2.5	2.5	2.4
b. From family rules	2.3	2.1*	2.1	2.0
5. Viewing flexibility				
a. Decision making	3.8	3.9	3.9	3.9
b. Technical control	2.5	2.5	2.5	2.4

*The asterisked comparisons indicate statistically significant differences at $p < .05$, using two-tailed tests.
a. The higher the mean, the more frequent the behavior.

activity. The viewing flexibility indexes also did not differentiate these groups.

Gratifications sought and obtained. Having a VCR did not alter the gratifications sought or obtained in a consistent or significant manner. The gratifications do parse out, however, as indicators of what television is used for. Entertainment and diversion clearly top the set of gratifications sought and obtained with surveillance close behind. Television is used minimally for informational guidance and for parasocial identity.

VCR behaviors. The final two tables look at what is done with the VCR in households in which they are fairly new additions and in homes where they have been available for more than two years. Adolescents more acclimated to the VCR did more planning of the shows they were going to tape while watching or not watching television concurrently. However, these same youngsters did *less* zapping of commercials while

Table 5.5 VCR Behaviors by New and Old Access

Nature of Use[a]	New Access	Old Access
1. Prior Planning	3.4	3.7*
2. Zapping	3.6	3.4*
3. Hours Viewed		
Mon-Fri taping	2.4 hrs.	2.8
Sat-Sun taping	2.6	2.8
Friends' taping	2.9	2.9
Home rentals	1.2	1.4
Friends' rentals	.7	.9
4. Show types preferred		
Soap operas	1.5	1.7
Police/detective	2.0	2.2
Sitcoms	2.8	3.0
Dramatic series	1.8	1.9
Movies or miniseries	2.9	3.3*
Sports	2.3	2.5
Music videos	2.0	2.2
Advice columns	1.1	1.2
News	1.4	1.2
Public affairs	1.3	1.3

*The asterisked comparisons indicate statistically significant differences at $p < .05$, using two-tailed tests.
a. The higher the mean, the more frequent the behavior or the preference.

taping or zipping through commercials during playback. There are no reported differences in Table 5.5 in how much time is spent with the VCR, either in terms of taping or playing back prerecorded videos. Somewhat more weekday time, however, was given to use of the VCR in the older VCR homes ($p < .09$). Time devoted to watching videos taped at home or at a friend's home, however, is more than double that of watching rental videos. In all, this sample of adolescents claims to spend about 10 hours a week watching shows on their VCR or that of their friends. Finally, no differences were found in show types preferred between the new and old VCR youth. Sitcoms and movies topped the preferred list for both groups.

Control of the VCR. The adolescents in this sample said that they are consistently in control of the VCR. They claim they can decide when they want to tape something, how many shows they want to tape, which

Table 5.6 VCR Control and Mediation by New and Old Access

	New Access	Old Access
VCR use control[a]		
a. Control over when	3.9	4.1*
b. Control over what	3.7	3.9
c. Control over how many	3.4	3.7*
d. Zipping	3.5	3.8*
Mediation		
a. From parents	2.5	2.3
b. From family rules	2.1	1.9

*The asterisked comparisons indicate statistically significant differences at $p < .05$, using two-tailed tests.
a. The higher the mean the more frequent the reported control or mediation.

shows they want to tape, and whether or not to zip and zap commercials during playback and taping.

As to the comparisons of interest, the adolescents who have had VCRs in their homes longer are able to perform each of these behaviors more regularly, including a strong tendency to have greater control over what is taped ($p < .08$). Mediation from parents or established rules do not differentiate the new and old VCR respondents, although the latter tended to report fewer family rules ($p < .10$).

DISCUSSION

When a communication technology has diffused through half the population, and especially when it has diffused through three-fourths of an adolescent sample, who remains? Are they hard-core nonadopters, laggards waiting to be sure they can make the best buy, firm rejectors, or the truly impoverished? Our data rule out only the last plausible explanation. Although SES attributes continue to offer strong and consistent differences between those who have and do not have a VCR, the nonowners in this study are far from poor. They have two or three cars, they drive to three or four bedroom homes equipped with two or three television sets, and a majority have a personal computer.

Thus, their nonadoption of video warrants a different perspective. A likely basis for their nonadoption may lie in a family or parental orientation that diminishes the need for even more television or a form of television that is more out of the control of their parents.

More perplexing are the findings about the prevalence of VCRs across homes with different parent structures, in particular greater penetration in homes with both original parents. By contrast, Greenberg and Heeter (1987) found the presence of both original parents equally in VCR and non-VCR homes three years earlier in a similar population, while finding lower levels of penetration in homes where one parent stayed home and the other worked. The latter finding suggested that the nontraditional family would more rapidly adopt; now, with penetration perhaps peaking, the traditional parental arrangement is a superior predictor. Clearly, it could be attributed to higher incomes in two parent versus one parent households. But VCRs are and have been available for $150 to $200 and that is unlikely to be an inhibiting price factor. Subsequently, more careful examination of who constitutes the parent group *in combination with* their employment status may provide better understanding of the diffusion of the next home entertainment communication technology, or at least its earlier adopters.

Most striking in this study is the magnitude of the prevalence of all forms of electronic gadgets among these adolescents. This suggests the continuing formation of a home video culture that now includes personal computers, video cameras, and CD players as supplements to both the old media and the extensive collection of new gadgets among the respondents' personal possessions. Recall that each item was possessed by more than one-half of the adolescents; if there were newer, less prevalent items being adopted, the investigators failed to identify them. In fact, of the seven items asked about, *each* adolescent possessed an average of five of them. The media environment is truly an abundant one for these adolescents, even more abundant than determined by this study, e.g., they were not asked to indicate how many radios they possessed, what software and how much, or the extent of their book, tape, record, disk collections. One might wish to determine if there is a "media aficionado" characteristic in adolescent subgroups. And even with the overall pervasiveness of these items, possession still discriminated those who already had VCRs from those who didn't and *continued* to distinguish the original VCR users from the more recent ones.

Despite the abundance of media in this teenage environment, availability did not translate into use. Estimated time spent with both print and electronic media was remarkably similar across all groups compared. While amount of time spent with television did not differ by VCR and non-VCR adolescents, it is not clear from the data whether asking VCR owners for television time, without carefully denoting that they should omit from their calculations the time spent watching television through a VCR, does or does not include that form of television. When asked subsequently to report time spent with VCRs, these respondents added another two hours a day to the total. This suggests that media-rich adolescents spread their time across the many options available to them, rather than either increasing their concentration on any one of them or increasing their overall media time. Thus, time formerly given to broadcast and/or cable television is now shared with video.

Our study also replicates other work which indicates there is one particular difference in content preferences in homes with more advanced technology—a preference for more sexual content in media messages. Respondents with VCRs claimed more frequent viewing of R-rated movies from their pay channels, and likely, taping of them as well. Without a panel study, it is not possible to separate those who sought cable and VCRs in order to more easily access this content, as Roe (1987) found among Swedish adolescents, from those whose appetites have been intensified after finding such content available. Whichever route it takes, this increased exposure to sex content, in a context of minimal parental mediation across this broad age group, provides an opportunity for research on its impact.

The lack of significant differences among the viewing process behaviors are disappointing. How television is watched did not seem greatly affected by the presence of a VCR intervening, although we know from earlier studies (Heeter & Greenberg, 1988) that television watching is considerably different in cable and noncable TV environments. Cable induces more planning, more channel switching, more multichannel viewing, among other changes. By contrast the VCR primarily promotes zipping and zapping through commercials. One needs to examine these behaviors in a more appropriately designed study if the goal is to determine which viewing behavior changes are a function of having cable and/or a VCR, i.e., at least a two-by-two matrix of adolescents with and without cable *and* with and without a VCR in the home. Extending that design to include some measure of length of time with the technology, as has been done in this study, is one way to understand long-term changes attributable to technological innovations.

The importance of the time dimension in understanding the impact of new technology on the mass communication process is exemplified by the obtained results for VCR behaviors among the new and older users. More planning of what to do with the VCR, more perceived control over when, what and how much to tape, and when to zip and zap may be attributed to the experience accumulated after the initial period of fascination and experimentation with a new piece of technology; from this phase, the viewer may gain mastery over what it is possible to control and then choose what he or she wishes to continue to do. However, why more experienced VCR adolescents do less excising of commercials (while doing more than non-VCR owners) remains unresolved. It should be noted that the differences between old and new users is not explainable by the age of the respondents; there was no relationship between year in school (seventh and tenth grades) and whether they had a VCR for more or less than two years in their home.

The VCR is now added to an increasingly rich media environment— rich in alternative technologies for accessing entertainment and information and rich in alternative sources of entertainment and information content. How do these pieces of the media puzzle fit together in the lives of adolescents, or adults, for that matter? To study these technologies one at a time, as has largely been done, leads immediately to questions of their interaction with other media. Already noted is the need to examine the interaction of VCR use with and without cable. Studies of electronic media among adolescents tend to ignore depth studies of print media, settling for one or two questions about time commitment and often ignoring content preferences altogether. A more holistic approach is overdue, one which logs the composite media behaviors of adolescents, attempts to understand their content choices within alternative media, looks to how they process content in different media, and assesses the media's social impacts in an increasingly complex media environment. Complex at least to the researcher, if not to the adolescent who may be handling all this and more with a teenager's blasé shrug.

REFERENCES

Atkin, D., Heeter, C., & Baldwin, T. (1987). *Parental mediation: A comparison of pay, basic and noncable homes.* Paper presented at the meeting of the International Communication Association, Montreal.

Bantz, C. R. (1982). Exploring uses and gratifications: A comparison of reported uses of television and reported uses of favorite program type. *Communication Research, 9,* 352-379.

Blumler, J. G. (1979). The role of theory in uses and gratifications studies. *Communication Research, 6,* 9-36.

Brown, J. D., Bauman, K. E., Lentz, G. M., & Koch, G. G. (1987). *Young adolescents' use of radio and television in the 1980s.* Paper presented at the meeting of the International Communication Association, Montreal.

Greenberg, B. S. (1974). Gratifications of television viewing and their correlates for British children. In J. G. Blumler & E. Katz (Eds.), *The uses of mass communications: Current perspectives on gratifications research.* Beverly Hills, CA: Sage.

Greenberg, B. S., & Heeter, C. J. (1987). VCRs and young people: The picture at 39% penetration. *American Behavioral Scientist, 30* (5), 509-521.

Haefner, M. J., Hunter, L. S., & Wartella, E. (1986). *Parents, children and new media: Expectations, attitudes and use.* Paper presented at the meeting of the International Communication Association, Chicago.

Hamburg, B. (1982). Theses and variations of adolescence. In M. Schwartz (Ed.), *TV & teens: Experts look at the issues* (pp. 2-99). Menlo Park, CA: Addison-Wesley.

Heeter, C., & Greenberg, B. S. (1988). *Cableviewing.* Norwood, NJ: Ablex.

Himmelweit, H., & Swift, B. (1976). Continuities and discontinuities in media usage and taste: A longitudinal study. *Journal of Social Issues, 32*(3), 133-156.

Horton, D., & Wohl, R. (1956). Mass communication and para-social interaction. *Psychiatry, 19,* 215-229.

Hughes, C. E., & Dobrow, J. R. (1988). *The VCR and the adolescent: Patterns of use.* Paper presented at the meeting of the International Communication Association, New Orleans.

Johnsson-Smaragdi, U., & Roe, K. (1986). *Teenagers in the new media world: Video recorders, videogames and home computers.* Lund Research papers in the *Sociology of Communication, 2.* University of Lund, Sweden.

Levy, M. R. (1978). The audience experience with TV news. *Journalism Monographs,* No. 55.

Levy, M. R. (1979). Watching TV news as parasocial interaction. *Journal of Broadcasting, 23*(1), 69-80.

Levy, M. R. (1987). VCR use and the concept of audience activity. *Communication Quarterly, 35*(3), 267-275.

Levy, M. R., & Fink, E. (1984). Home video recorders and the transcience of television broadcasts. *Journal of Communication, 34*(2), 56-71.

Lin, C. A. (1987). *Adolescents in the multimedia environment: Their media use activities and gratifications.* Unpublished doctoral dissertation, Michigan State University.

Lin, C. A. (1988). *Assessing the impact of the evolution of home video culture.* Paper presented at the meeting of the Association for Education in Journalism and Mass Communication, New Orleans.

Lin, C. A., & Atkin, D. (1988). *Parental mediation and adolescent uses of television and VCRs.* Paper presented at the meeting of the International Communication Association, New Orleans.

McQuail, D., Blumler, J. G., & Brown, J. R. (1972). The television audience: a revised perspective. In D. McQuail (Ed.), *Sociology of mass communications.* Harmondsworth, UK: Penguin.

Medrich, E. A., Rozien, J., Rubin, V., & Buckley, S. (1985). *The serious business of growing up.* Berkeley, CA: University of California Press.

Roe, K. (1983a). *The influence of video technology in adolescence.* Media panel report No. 27, University of Lund.

Roe, K. (1983b). *Mass media and adolescent schooling: Conflict or coexistence?* Stockholm: Almquist & Wiskell International.

Roe, K. (1987). Adolescents' video use: A structural-cultural approach. *American Behavioral Scientist, 30*(5), 523-532.

Rubin, A. (1983). Television uses and gratifications: The interactions of viewing patterns and motivations. *Journal of Broadcasting, 27*(1), 37-51.

Rubin, A. M., & Bantz, C. R. (1987). Utility of videocassette recorders. *American Behavioral Scientist, 30*(5), 471-485.

Schramm, W., Lyle, J., & Parker, E. B. (1961). *Television in the lives of our children.* Standford, CA: Standford University Press.

Teitelbaum, H. (1988, June 27). Usage of VCRs by kids tops adults, AGB Says. *Multichannel News,* p. 15.

6

Social and Psychological Antecedents of VCR Use

ALAN M. RUBIN
and
REBECCA B. RUBIN

The videocassette recorder (VCR) became a socially significant communication technology as it moved into the home in the 1980s. VCRs provide expanded content and context options over traditional media. They accentuate choice, involvement, and control, and highlight the active and interactive nature of personal and mediated communication. Research about the nature and impact of VCRs, however, is limited even though over 60% of all U.S. households now have VCRs.

We take a uses and gratifications (U&G) approach to studying VCRs. U&G is based on the tenet that social and psychological factors influence people's motives to communicate, choices of communication alternatives, behavior, and communication effects. An underlying U&G assumption is that people choose to communicate purposely to satisfy felt needs; this behavior produces gratifications. People are seen as variably active communicators.

U&G is especially appropriate for studying VCRs, which invite active audience participation by allowing greater control over viewing choices than traditional media. According to U&G, we must first understand communication motivation to interpret communication effects. To understand motivation, we must assess social and psychological antecedents, and personal and mediated functional alternatives. This investigation extends research into the interface of personal and

Copyright © 1989 by Alan M. Rubin and Rebecca B. Rubin

mediated communication, and synthesizes the study of communication motivation and outcomes.

The VCR is an evolving technology with an increasing impact on communication in the home. Our goal is to explain social and psychological antecedents of VCR use and to consider the interplay of personal and mediated communication. We test a model by which we expect VCR motives to complement interpersonal communication motives, which are influenced by life position, which is affected by personality and demographic factors.

Motives for Using VCRs

Harvey and Rothe (1985-1986) identified six basic reasons to use VCRs: to zap commercials, to time shift, to establish an environment for children, to increase viewing choices, to increase noncommercial viewing by building a library of programs, and to fast view by zipping through programs. Time shifting and increasing viewing choices were the most important uses. They also described a socializing function as new VCR owners invited others over to watch special events.

Rubin and Bantz (1987, 1988) employed a U&G framework to examine VCR motives. In order of importance, they found eight interrelated motives for using VCRs: movie rental, time shifting, library storage, socializing, music video viewing, critical viewing, exercise tape viewing, and child viewing. They concluded that VCRs highlight the role played by individual differences in communication. For example, younger persons often used music videos, exercise tapes, and movies to shift time, view critically, and socialize.

Rubin and Bantz (1987, 1988) proposed that VCR use is active behavior that complements and extends other modes of communication. VCR use is a functional alternative to interpersonal communication. They argued that we need to examine VCR use in relation to interpersonal communication and to examine the social and psychological antecedents of VCR use, including how an individual's life position and sense of life control affect VCR use.

Investigators have observed that communication behavior affects why we use VCRs. For example, Dobrow (1987) found that heavier television viewers used VCRs to concentrate their viewing on favorite or specific types of programs. Lighter television viewers used VCRs to provide greater viewing diversity. We also would expect that different motives for using VCRs should help explain frequency of VCR use.

Personal and Mediated Communication

Although U&G typically has been used in mass communication, we have proposed its relevance for interpersonal communication (Rubin & Rubin, 1985). Interpersonal channels are need-gratification alternatives, which may be coequal to mediated channels. For example, one can gratify companionship needs by conversing with a friend or by listening to talk radio. Social and psychological antecedents affect both interpersonal and media motives. They also influence the choices of alternatives for gratifying felt needs and how adequate these choices are.

The social nature of VCR use requires consideration of interpersonal and social interaction. Schoenbach and Hackford (1987) found that West German video households had more leisure-time activities, and that nonusers of VCRs were not more physically active that VCR users. In a study of interpersonal communication and media consumption in Saudi Arabia, Al-Attibi (1986) found that interpersonal communication fulfilled adolescents' affective, entertainment, and escape needs, whereas the media gratified information needs.

Although little attention has been given to motives for interpersonal communication, researchers have suggested that interpersonal communication is goal-directed and that people communicate with one another to satisfy needs for inclusion, affection, and control (Schutz, 1966). A recent study of interpersonal communication motives identified three additional needs: pleasure, escape, and relaxation (Rubin, Perse, & Barbato, 1988). The development of this interpersonal-motives measure allows us to examine connections between interpersonal and VCR motives.

Life Position and Communication

Life position affects both personal and mediated communication. We have conceptualized life position as "contextual age," a constellation of social, psychological, economic, health, and communication indicators of age (A. Rubin & R. Rubin, 1982). We developed contextual age as an alternative to chronological age because the latter improperly assumes homogeneity along such life-position dimensions as life satisfaction, mobility, and interaction (Rubin & Rubin, 1986). Researchers have suggested links among life-position dimensions such as health and social activity (Graney & Zimmerman, 1980), health and life satisfaction (Maddox & Eisdorfer, 1962), and social interaction and life satisfaction (Duff & Hong, 1982; Havighurst, Neugarten, & Tobin, 1968; Tesch, Whitbourne, & Nehrke, 1981).

Life position affects how people use media (A. Rubin & R. Rubin, 1982; R. Rubin & A. Rubin, 1982). For example, older elderly, who were less self-reliant and less interpersonally and socially active, felt that television was important in their lives. Those low in life satisfaction and economic security tended to watch television for companionship and escape. Wenner (1976) found that life descriptors such as household size related to viewing frequency and motivation for the elderly. Socially mobile elderly, for example, used television to avoid social contact with others, but socially isolated elderly used television content to facilitate interaction with others. Rubin (1986) found that the less mobile, healthy, active, and life satisfied became dependent on television to fulfill communication needs. Mediated communication, then, can become a substitute for interpersonal interaction. These findings emphasize the role life position plays in the interface of personal and mediated communication.

Locus of Control

The notion of "control" also is important when addressing communication behavior. Rubin (1986) argued that: "We need to consider whether locus of control, alone or in combination with other factors, produces variations in motives for and consequences of using personal and mediated information channels" (p. 135). Locus of control affects behavior (Rotter, 1954). "Internals" feel they control events in their lives, whereas "externals" view life outcomes as dependent on luck, chance, or powerful others. Pointing to the research of Williams, Phillips, and Lum (1985) and Schoenbach and Hackforth (1987), Levy (1987) asked whether consumers use VCRs for control.

Locus of control commonly refers to a person's mastery of his or her environment and life. Locus of control is consistent with U&G's active audience concept. Active audience members seek to gratify their needs and control their actions. According to Brenders (1987), "internals should be motivated to seek out, exert greater effort in, and derive greater satisfaction from situations allowing personal control" (p. 96).

Klippel and Sweeney (1974) suggested that externally controlled elderly would use less formal information sources.

Locus of control relates to cognitive activity. Lefcourt (1982) summarized several studies suggesting that internally controlled persons pay close attention to cues that allow for uncertainty reduction and task accomplishments. Internals were more variable than externals in their attentiveness. For tasks requiring less skill or thought, internals acted more impulsively, but took more time and effort with difficult tasks. According to Lefcourt, internals are more self-directed, whereas externals are other- or chance-directed and powerless, and would have a greater need for affiliation with others.

Demographic Characteristics

Demography affects life position and communication behavior. For example, we have found education to relate positively to mobility and economic security, and age to relate positively to economic security and life satisfaction but negatively to health and mobility (Rubin & Rubin, 1986). Demography also affects VCR use. When examining the social context of VCR viewing in Great Britain, Gunter and Levy (1987) found male/female differences and that most VCR playback was done alone. Rubin and Bantz (1987) identified age and gender differences in why VCRs were used. And, Dobrow (1987) linked VCR use with education, income, and ethnic background.

Age is one influence on how and why we use VCRs. Greenberg and Heeter (1987) found that VCR youth had lower self-evaluations and religious tendencies, but VCR youth among the 10th grade sample had positive self-evaluations and greater satisfaction with their peer relationships. Roe (1987) found that Swedish teens, especially low achievers in school, used VCRs in groups. The VCR, though, didn't contribute to low achievement. This research suggests that we should consider demography when examining antecedents of VCR use.

Model of VCR Use

In this study, then, we employed a U&G orientation to examine antecedents of VCR use. Primarily, we considered the model presented in Figure 6.1. The model outlines an expected sequence to the social and psychological antecedents of VCR use: demography and locus of

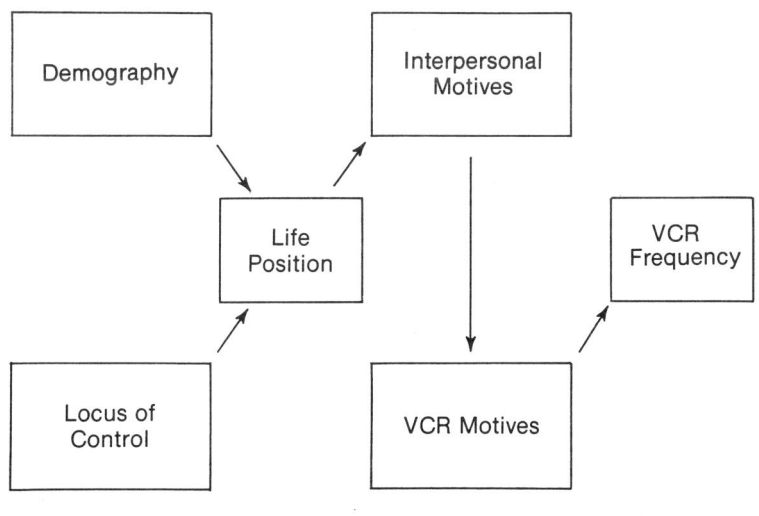

Figure 6.1. Antecedents of VCR Use

control influence a person's social and psychological well-being or life position, life position affects motives for interpersonal communication, interpersonal motives lead to VCR motives, and VCR motives affect the frequency of VCR use.

METHOD

Sample and Procedures

Similar to past survey-research studies (e.g., Rubin & Bantz, 1987; Rubin et al., 1988), we included a wide range of people in the sample by using purposive, quota sampling. Students enrolled in two undergraduate communication research classes at Kent State University were given specific age and gender sampling quotas. We trained these assistants in data collection and research ethics, and instructed them to solicit one male and one female of various educational backgrounds from each of four age groups to represent the general population: 18-34, 35-49, 50-64, and 65 years and over. Questionnaires were anonymous and individually self-administered. Because they were completed over spring break in 1988, assistants were able to recruit volunteers external to the campus, mostly from their home towns. A total of 428 completed questionnaires were returned; 299 of these respondents (69.9%) owned or used a VCR at home. The latter group constituted the sample for this study.

Respondents in the VCR sample ranged in age from 18 to 75 ($M = 42.99$, $SD = 15.39$); 52.8% were male (0 = male, 1 = female), 67.6% were presently married, and 79.4% were employed outside the home. The average household size was 3.17 persons ($SD = 1.20$). Most respondents were at least high school graduates ($M = 3.98$, $SD = 1.11$, where 1 = elementary school, 2 = some high school, 3 = high school graduate, 4 = some college, 5 = college graduate, and 6 = advanced degree). Respondents were mainly from the midwestern region of the U.S., with 84.8% of the VCR sample from Ohio.

Measurement

Locus of control. We measured locus of control (LOC) with Levenson's (1974) scale. The scale uses a Likert format rather than forced-choice responses as found in Rotter's (1966) scale, and is not prone to social desirability (Levenson, 1974) or self-conscious (Mikawa, Nordin, & Eyman, 1986) biases. Respondents reported their agreement with 12 statements (1 = strongly disagree, 5 = strongly agree). Four summed and averaged items represented each of three LOC dimensions: *Powerful Others Control* ($M = 2.43$, $SD = .69$, alpha = .70), *Internal Control* ($M = 3.69$, $SD = .57$, alpha = .64), and *Chance Control* ($M = 2.46$, $SD = .66$, alpha = .66).[1]

Contextual age. Contextual age (CA) reflects life-position rather than just chronological age (Rubin & Rubin, 1986). Respondents stated their agreement with 18 statements (1 = strongly disagree, 5 = strongly agree). Three summed and averaged items were used for each of four CA dimensions: *Physical Health* ($M = 3.66$, $SD = .77$, alpha = .62), *Mobility* ($M = 4.11$, $SD = .85$, alpha = .65), *Life Satisfaction* ($M = 3.66$, $SD = .71$, alpha = .72), and *Economic Security* ($M = 3.10$, $SD = .92$, alpha = .81). To improve reliability, we combined the six interpersonal interaction and social activity items into a fifth CA dimension of *Interaction* ($M = 3.31$, $SD = .63$, alpha = .62).[2]

Interpersonal communication motives. We used the Interpersonal Communication Motives scale to measure interpersonal (IP) motives (Rubin et al., 1988). Respondents reported how much each of 18 statements was like their own reasons for talking to people (1 = not at all, 5 = exactly). Three summed and averaged items were used for each of six IP motives: *Relaxation* ($M = 3.02$, $SD = .85$, alpha = .75), *Pleasure* ($M = 3.48$, $SD = .87$, alpha = .80), *Control* ($M = 2.47$, $SD = .88$, alpha = .69), *Inclusion* ($M = 2.99$, $SD = .97$, alpha = .77), *Affection* ($M = 3.75$, $SD = .76$, alpha = .74), and *Escape* ($M = 2.25$, $SD = .87$, alpha = .66).[3]

VCR motives. We adapted earlier measures of television viewing motives (Rubin, 1983) and VCR use (Rubin & Bantz, 1987) to assess motives for using a VCR. Respondents stated how much each of 22 reasons for using a VCR at home was like their own reasons (1 = not at all, 5 = exactly). Given the exploratory level of this analysis compared with the previous established scales, responses were subjected to principal components factor analysis with iterations and oblique rotation (SPSS, 1988). The factor solution explained 63.9% of the total variance. We required factors to have minimum eigenvalues of 1.0 and at least two items with .50 minimum primary loadings and no secondary loadings greater than .30 on retained factors. There were five factors (see Table 6.1). VCR motives were computed by summing and averaging responses to the acceptable items on each factor.

Factor 1, *Library Storage* ($M = 2.65$, $SD = .93$, alpha = .81), was using the VCR to save, store, and retrieve tapes for repeated viewing. It had an eigenvalue of 6.67 and explained 11.2% of the total variance after rotation. Factor 2, *Social Interaction* ($M = 2.45$, $SD = .95$, alpha = .78), was using the VCR to entertain and to interact socially with others. It had an eigenvalue of 1.97 and explained 10.0% of the total variance. Factor 3, *Freedom of Choice* ($M = 3.72$, $SD = .73$, alpha = .68), was using the VCR for convenient and cost-effective choice. It had an eigenvalue of 1.83 and explained 11.9% of the total variance. Factor 4, *Learning* ($M = 2.55$, $SD = .91$, alpha = .75), was using the VCR to acquire information. It had an eigenvalue of 1.44 and explained 8.5% of the total variance. Factor 5, *Time Shifting* ($M = 3.47$, $SD = 1.02$, alpha = .77), was using the VCR to shift, recover, and save time. It had an eigenvalue of 1.05 and explained 10.2% of the total variance. Freedom of choice and time shifting were the two most salient VCR motives.

Table 6.1 Primary Factor Loadings of VCR Motives

VCR Motive Items "I Use a VCR at Home Because..."	VCR Motive Factors				
	LIBR	SOCL	FREE	LERN	TIME
FACTOR 1: LIBRARY STORAGE					
I like to have movies or programs that can be viewed many times	.84	.11	.08	-.18	-.04
I like to save tapes for repeated viewing	.76	.04	-.01	.00	.22
I want to keep a permanent copy of the program	.68	.10	-.06	.06	.15
I want to re-watch a program and review it critically	.61	-.02	-.02	.22	.04
FACTOR 2: SOCIAL INTERACTION					
I want to entertain others at parties	.08	.90	-.12	-.01	.05
I want to entertain people who come over	.13	.87	.00	.02	-.06
I can be with others who are watching	-.16	.55	.15	.20	.05
FACTOR 3: FREEDOM OF CHOICE					
It gives you more choice over what to watch	-.04	.02	.77	.12	.05
It doesn't cost much money	-.09	.09	.58	-.14	.10
It provides options to regularly scheduled TV programs	.06	-.12	.55	.24	.18
I like the freedom to set my own schedule	-.02	.21	.51	.04	.23
FACTOR 4: LEARNING					
I can learn something	.01	.05	.04	.85	-.04
I can learn things I haven't done before	-.03	.10	-.05	.83	.02
FACTOR 5: TIME SHIFTING					
I want the convenience of recording programs and watching them later	.13	-.06	.07	-.02	.80
When I view later, I can skip through commercials	.07	.13	-.02	-.11	.76
I am busy doing something when a program is on the air	.04	.02	.04	.14	.72

VCR frequency. We asked respondents to estimate about how many hours of television they usually watch each day, how many days each week they usually use a VCR at home, and, on those days they use a VCR, how many hours each day they usually use the VCR to watch or record programs or tapes. Television viewing averaged 3.24 hours ($SD = 2.03$) each day. Daily VCR use averaged 2.51 hours ($SD = 1.64$) on an average 2.29 days ($SD = 1.77$) each week. We multiplied the days per week and hours per day of VCR use to create an index of weekly VCR Frequency ($M = 6.63$ hours, $SD = 8.49$).

RESULTS

Correlates of VCR Motives and Frequency

We first looked at the partial correlates of VCR motives and frequency. Several conclusions are evident from data in Table 6.2. First, VCR motives were interrelated. The strongest associations were between using the VCR for: library storage and both time shifting and social interaction, and freedom of choice and time shifting. Second, VCR motives correlated more with interpersonal motives than with other antecedents. VCR motive correlates were: (a) library storage with IP inclusion and control, CA immobility, and household size; (b) social interaction with IP inclusion, escape, affection, relaxation, and pleasure, CA immobility and interaction, younger age, and LOC chance and powerful others control; (c) freedom of choice with IP inclusion, relaxation, pleasure, affection, and control, and LOC internal control; (d) learning with IP control and affection; and (e) time shifting with IP control and inclusion, LOC chance control, and CA unhealthy. Third, VCR frequency of use was linked more to motives for using VCRs than to other variables. VCR frequency correlates were: library storage, freedom of choice, time shifting, and social interaction VCR motives; less education; LOC external, powerful others, and chance control; and IP control.

Predictors of VCR Motives and Frequency

Next, we asked which antecedent variables best explained VCR motives. We regressed each VCR motive on blocks of antecedent variables, which were entered into the equation based on the conceptual model: (a) demography, (b) locus of control, (c) contextual age, and (d) interpersonal motives. Demography was entered on a separate first step to control for demographics prior to the entry of locus of control and the other variables. To locate predictors of VCR frequency, we also regressed frequency on the same blocks of antecedent variables with the addition of VCR motives on a fifth step. Table 6.3 summarizes the hierarchical multiple regression analyses. Five of the six equations were significant: library storage, social interaction, freedom of choice, time shifting, and VCR frequency.

Library storage. The demographics explained 4.1% of the library storage variance ($R^2 = .04$, F change $= 3.14$, $p < .20$) on step 1. House-

Table 6.2 Partial Correlates of VCR Motives and Frequency

Correlates	LIBR	SOCL	FREE	LERN	TIME	FREQ
Demography						
Age	−.09	−.26***	−.07	.07	−.11	−.06
Gender	−.07	.06	−.04	.01	−.04	.02
Education	.01	−.04	.04	.09	−.03	−.21***
Household size	.12*	.02	.05	.10	.09	.11
Locus of Control						
Powerful others	.03	.15**	.04	.08	.10	.13*
Internal control	−.04	−.03	.12*	.08	−.03	−.16**
Chance control	.06	.17**	.07	.02	.15*	.11*
Contextual Age						
Physical health	−.04	−.07	−.08	−.01	−.13*	−.09
Interaction	.11	.23***	.05	.10	.05	−.02
Mobility	−.13*	−.34***	−.02	−.06	−.08	−.05
Life satisfaction	.03	.03	.05	.04	−.11	−.07
Economic security	−.01	−.03	−.08	.08	.02	−.06
Interpersonal Motives						
Relaxation	.08	.20***	.19***	.07	.10	−.06
Pleasure	.09	.20***	.18**	.08	.09	.02
Control	.16**	.07	.12*	.13*	.20***	.14*
Inclusion	.20***	.37***	.23***	−.01	.14*	.06
Affection	.10	.21***	.16**	.12*	.02	−.02
Escape	.11	.28***	.06	−.01	.06	.02
VCR Motives						
Library storage	—					
Social interaction	.42***	—				
Freedom of choice	.35***	.36***	—			
Learning	.34***	.34***	.27***	—		
Time shifting	.46***	.19***	.44***	.32***	—	
VCR Frequency	.28***	.18**	.24***	.11	.22***	—

NOTE: Coefficients are fourth-order partial correlations controlling for all four demographics, except for demographic coefficients which are third-order partial correlations controlling for the other three demographics.
* $p < .05$, ** $p < .01$, *** $p < .001$ (two-tailed).

hold size was a positive predictor. On step 2, locus of control explained less than 1% additional variance ($R^2 = .04$, F change = .20, $p = .89$). Contextual age accounted for 2.6% further variance on the third step ($R^2 = .07$, F change = 1.58, $p = .17$). Gender emerged as a negative predictor. On the final step, the interpersonal motives explained 4.8% more variance ($R^2 = .12$, F change = 2.49, $p < .03$). IP control was a positive predictor. Significant final predictors of the library storage VCR motive were: communicating interpersonally for control, larger

Table 6.3 Hierarchical Regression: Predicting VCR Motives and Frequency

	LIBR final b	SOCL final b	FREE final b	LERN final b	TIME final b	FREQ final b
STEP 1: DEMOGRAPHY						
Age	−.10	−.22***	−.06	.04	−.15*	.01
Gender	−.12*	−.09	−.06	.02	−.05	.06
Education	.03	−.01	.05	.11	−.01	−.21***
Household size	.15*	.03	.02	.09	.10	.06
STEP 2: LOCUS OF CONTROL						
Powerful others	−.09	.03	.02	.07	−.05	.05
Internal control	−.07	.04	.15*	.09	.00	−.15*
Chance control	.02	.00	.05	.00	.11	.02
STEP 3: CONTEXTUAL AGE						
Physical health	−.05	−.03	−.09	−.07	−.10	−.04
Interaction	.11	.17**	.01	.08	.11	.03
Mobility	−.08	−.26***	.00	−.06	.00	.03
Life satisfaction	.02	.01	.04	−.06	−.15*	.00
Economic security	.02	−.01	−.12	.06	.08	−.02
STEP 4: IP MOTIVES						
Relaxation	−.01	−.03	.07	.07	.06	−.09
Pleasure	−.01	.00	.08	.02	.06	.08
Control	.15*	−.02	.09	.14*	.23***	.10
Inclusion	.12	.21**	.13	−.13	.02	−.02
Affection	.05	.15*	.06	.12	−.02	−.08
Escape	.02	.15*	−.07	−.03	−.05	−.06
STEP 5: VCR MOTIVES[a]						
Library storage	—	—	—	—	—	.18**
Social interaction	—	—	—	—	—	.09
Freedom of choice	—	—	—	—	—	.15*
Learning	—	—	—	—	—	−.03
Time shifting	—	—	—	—	—	.04

NOTE: *Betas* are standardized *beta* weights when entered at Step 4 for VCR Motives or at Step 5 for VCR Frequency.
a. VCR Motives were entered only for VCR Frequency. * $p < .05$, ** $p < .01$, *** $p < .001$.
LIBR: $R = .34$, $R^2 = .12$, $F(18, 277) = 2.04$, $p < .01$
SOCL: $R = .56$, $R^2 = .31$, $F(18, 279) = 6.99$, $p < .001$
FREE: $R = .35$, $R^2 = .13$, $F(18, 279) = 2.23$, $p < .01$
LERN: $R = .29$, $R^2 = .08$, $F(18, 279) = 1.41$, $p = .12$
TIME: $R = .36$, $R^2 = .13$, $F(18, 279) = 2.27$, $p < .01$
FREQ: $R = .46$, $R^2 = .21$, $F(23, 270) = 3.14$, $p < .001$

household size, and male gender. The measures explained 11.7% of the library storage variance.

Social interaction. On step 1, the demographics explained 7.5% of the social interaction variance ($R^2 = .07$, F change = 5.92, $p < .001$). Age was a negative predictor. Locus of control explained 1.9% further variance on step 2 ($R^2 = .09$, F change = 2.01, $p = .11$). Contextual age accounted for 12.0% more variance on the third step ($R^2 = .21$, F change = 8.68, $p < .001$). Mobility was a negative predictor, and interaction was a positive predictor. The interpersonal motives explained 9.7% additional variance on the last step ($R^2 = .31$, F change = 6.58, $p < .001$). IP inclusion, escape, and affection were positive predictors. Significant final predictors of the social interaction VCR motive were: communicating interpersonally for inclusion, escape, and affection, CA interaction, CA immobility, and younger age. The measures explained 31.1% of the social interaction variance.

Freedom of choice. On the first step, the demographics accounted for 1.9% of the freedom of choice variance ($R^2 = .02$, F change = 1.40, $p = .23$). Locus of control explained 2.4% further variance on step 2 ($R^2 = .04$, F change = 2.40, $p = .07$). Internal control was a positive predictor. On step 3, contextual age also accounted for 2.4% additional variance ($R^2 = .07$, F change = 1.46, $p = .20$). The interpersonal motives explained 5.9% more variance on the final step ($R^2 = .13$, F change = 3.16, $p < .01$). At the conclusion of the analysis, the only significant predictor of the freedom of choice VCR motive was LOC internal control. The measures explained 12.6% of the freedom of choice variance.

Learning. On step 1, the demographics accounted for 1.7% of the learning variance ($R^2 = .02$, F change = 1.27, $p = .28$). Locus of control also explained 1.7% more variance on the second step ($R^2 = .03$, F change = 1.68, $p = .17$). Contextual age accounted for 1.3% further variance on step 3 ($R^2 = .05$, F change = .78, $p = .56$). The interpersonal motives explained 3.7% more variance on the final step ($R^2 = .08$, F change = 1.86, $p = .09$). IP control was a positive predictor. Although communicating interpersonally for control predicted the learning VCR motive, the regression equation was not significant. The predictors explained only 8.3% of the learning variance.

Time shifting. The demographics accounted for 3.0% of the time shifting variance on step 1 ($R^2 = .03$, F change = 2.24, $p = .07$). Age was a negative predictor. On step 2, locus of control explained 1.6% further variance ($R^2 = .05$, F change = 1.63, $p = .18$). Contextual age accounted for 2.9% more variance on the third step ($R^2 = .07$, F change = 1.78, $p = .12$). The interpersonal motives explained 5.3%

further variance on the last step ($R^2 = .13$, F change = 2.84, $p < .02$). CA life satisfaction emerged as a negative predictor, and IP control was a positive predictor. Significant final predictors of the time shifting VCR motives were: communicating interpersonally for control, CA life dissatisfaction, and younger age. The measures explained 12.8% of the time shifting variance.

VCR frequency. On step 1, the demographics explained 5.7% of the VCR frequency variance ($R^2 = .06$, F change = 4.38, $p < .01$). Education was a negative predictor. Locus of control accounted for 3.2% more variance on the second step ($R^2 = .09$, F change = 3.36, $p < .02$). Internal control was a negative predictor. Contextual age accounted for less than 1% additional variance on step 3 ($R^2 = .10$, F change = .40, $p = .85$). On step 4, the interpersonal motives explained 2.6% further variance ($R^2 = .12$, F change = 1.36, $p = .23$). The VCR motives accounted for 8.9% more variance on the final step ($R^2 = .21$, F change = 6.11, $p < .001$). VCR library storage and freedom of choice motives were positive predictors of VCR frequency. Significant final predictors of VCR frequency were: VCR library storage and freedom of choice motivation, LOC external control, and less education. The measures explained 21.1% of the VCR frequency variance.

Multivariate Relationships: VCR Use and Antecedents

In the last stage of the analysis, we used canonical correlation to assess the multivariate relationships among the set of antecedents to VCR use and the set of VCR motives and frequency. We included only those antecedents that were significant contributors in the regressions. Interpretation focused on canonical loadings of .30 or greater because smaller loadings may be unstable (Lambert & Durand, 1975). Canonical loadings are largely free of the effects of multicollinearity. The analysis identified two significant roots.

The first canonical root ($R_c = .58$, *lambda* = .47, $p < .001$) explained 33.1% of the common variance between the variates. Set 1 included positive correlations among IP inclusion (.60), IP escape (.50), CA interaction (.40), and IP affection (.31), and negative associations between these antecedents and both age (-.54) and CA mobility (-.48). The primary loading in set 2 was social interaction (.95), which correlated positively with library storage (.43), freedom of choice (.34), and time shifting (.30) VCR motives. Across the sets, younger, interactive, but less mobile persons, who communicated with others for reasons of inclusion, escape, and affection, used VCRs primarily for social interaction, and to store tapes, for convenient choice, and to time shift.

The second canonical root (R_c = *.39, lambda* = .70, p < .001) explained 14.9% of the common variance between the variates. Set 1 contained positive correlations among education (-.59), LOC internal control (-.59), and CA life satisfaction (-.41). The primary loading in set 2 was VCR frequency (.76), which correlated positively with time shifting (.32) and negatively with freedom of choice (-.30) VCR motives. Across the sets, less educated and externally controlled persons, who were less satisfied with their lives, were more frequent users of VCRs primarily to shift time rather than for convenient choice.

Discussion

The VCR is an evolutionary technology that is more than an appendage to television. VCRs allow us to use traditional channels of communication, such as television and interpersonal interaction, in different ways. For example, we use VCRs to interact socially with others. To socialize, we might first use VCRs to establish a library of programs and to reorder time. These motives speak to the purposeful and active uses of this medium, and to the links between personal and mediated communication.

Our results reinforce previous U&G notions. First, communication motives are interrelated. Similar to motives for television viewing and for interpersonal communication, motives for using the VCR are inherently related. The VCR is a convenient and economical mechanism for communication storage and retrieval. Convenience, choice-making ability, time restructuring, and social utility are central to VCR use. Such components reinforce earlier contentions that "VCR use is indeed active behavior" (Rubin & Bantz, 1988, p. 191).

Second, motives to communicate interpersonally predict motives for using a VCR. Communicating interpersonally for reasons of inclusion, affection, and escape, predicted social interaction motives for using the VCR. Those seeking companionship and interpersonal connection, as well as those wanting to alleviate boredom or put off doing something else, were likely to use the VCR to be with and entertain others. This, too, supports past research: VCR use links interpersonal and mass communication (Rubin & Bantz, 1987); and VCR use increases time spent with family and friends (Harvey & Rothe, 1986; Roe, 1987).

Social affiliation seems to be a salient interpersonal need whose gratification may be facilitated by using VCRs. In this study, younger, socially active, less mobile persons, who felt that other persons and chance occurrences influenced events in their lives, tended to have strong social uses for the VCR. This reinforces the argument that externals are other directed and need affiliation with others (Lefcourt, 1982). It also suggests that media such as VCRs can complement interpersonal communication channels in gratifying individual needs (Rubin & Rubin, 1985).

In addition, communicating interpersonally to achieve control (i.e., getting others to do something) predicted several VCR motives: time shifting, library storage, and learning. This supports Schutz's (1966) notion that control is an interpersonal communication need, yet further suggests that those who seek to achieve control can do so in both personal and mediated contexts. VCR use also enables one to master his or her environment. As Brenders (1987) noted, "one's belief or lack of belief in personal control profoundly influences his or her attitude toward the self and the environment" (p. 92).

This investigation, then, suggests that communication needs and motives are not restricted to interpersonal or mass communication contexts. For example, those who were motivated to communicate interpersonally for inclusion and affection, also used the VCR for social interaction purposes. Those who were motivated to communicate interpersonally to control others, also used the VCR to control their environment by storing tapes and time shifting. Communication needs appear to transcend communication channels or contexts. This reinforces the functional alternative notion of personal and mediated motives.

Third, there are social and psychological antecedents to communication motivation, which predicted VCR use. External locus of control and less education predicted frequency of VCR use. Perhaps the externally controlled use VCRs as a way to increase feelings of control in their lives. Males from larger households, who sought control in interpersonal communication, used VCRs to build a library of tapes. Again, those who seek control interpersonally can use the VCR to establish control of viewing options. Internal locus of control was the best predictor of the VCR freedom of choice motivation, which was a personal control factor emphasizing choice over what to watch, freedom to set one's own schedule, and multiple viewing options. Younger age and life dissatisfaction predicted using the VCR to shift time. And, younger age, restricted mobility, and social interaction predicted VCR use for social utility.

We found support, then, for U&G assumptions that there are social and psychological antecedents to communication motives and behavior, which must be examined and understood. In this study, social demography, psychological predispositions, and life position contributed to our understanding of VCR motives and behavior.

There are several directions for future research. Beyond motives, we need to consider content and consequences. We need to consider more directly what is being taped or replayed by VCR users and compare these choices with other media alternatives. Also, we need to examine the consequences of VCR use for social interaction and interpersonal relationships. How do VCRs alter family interaction and the home environment? How culture specific are such relationships? How do the increased possibilities of communication choices and control affect the uses and effects of other communication options, such as television and music? Do VCRs compensate for interpersonal communication deficiencies of the immobile or media dependent? Consequences also extend to societal structures. For example, how do VCRs affect network television programming and the future of network television? How do VCRs affect other communication organizations such as the music industry?

NOTES

1. The Locus of Control statements were: Powerful Others Control (I feel like what happens in my life is mostly determined by powerful people; People like myself have very little chance of protecting our personal interests when they conflict with those of strong pressure groups; My life is chiefly controlled by powerful others; Getting what I want requires pleasing those people above me); Internal Control (I am usually able to protect my personal interests; My life is determined by my own actions; When I make plans, I am almost certain to make them work; I can pretty much determine what will happen in my life); and Chance Control (To a great extent my life is controlled by accidental happenings; Often there is no chance of protecting my personal interest from bad luck happenings; When I get what I want, it's usually because I'm lucky; It's not always wise for me to plan too far ahead because many things turn out to be a matter of good or bad fortune).

2. The Contextual Age statements were: Physical Health (I usually feel in top-notch physical condition; Healthwise, I'm no worse off than anyone else my age; I have serious medical or health problems); Interpersonal Interaction (I get to see my friends as often as I would like; I spend enough time communicating with family or friends by telephone or mail; I have ample opportunity for conversations with other people); Mobility (I usually drive my own car or use the bus to get around; I have to rely on other people to take me places; I usually don't travel more than a few blocks from my home each day); Life

Satisfaction (I find a great deal of happiness in life; I've been very successful in achieving my aims or goals in life; I am very content and satisfied with my life); Social Activity (I often travel, vacation, or take trips with others; I often visit with friends, relatives, or neighbors in their homes; I often participate in games, sports, or activities with others); and Economic Security (I have no major financial worries; I have enough money to buy things I want, even if I don't really need them; I live quite comfortably now and have enough money to buy what I need or want). Negatively worded items were reverse coded for data analysis.

3. The Interpersonal Communication Motive statements were: Relaxation (because it relaxes me, because it allows me to unwind, because it's a pleasant rest); Pleasure (because it's exciting, to have a good time, because it's fun); Control (to get something I don't have, because I want someone to do something for me, to tell others what to do); Inclusion (because I need someone to talk to or be with, because it makes me feel less lonely, because I just need to talk about my problems sometimes); Affection (to let others know I care about their feelings, to help others, to thank them); and Escape (because I have nothing better to do, to get away from what I'm doing, to put off doing something I should be doing).

REFERENCES

Al-Attibi, A. A. M. (1986). Interpersonal communication competence and media consumption and needs among young adults in Saudi Arabia (Doctoral dissertation, Ohio State University, 1986). *Dissertation Abstracts International, 47*, 10A.

Brenders, D. A. (1987). Perceived control: Foundations and directions for communication research. In M. L. McLaughlin (Ed.), *Communication yearbook 10* (pp. 86-116). Newbury Park, CA: Sage.

Dobrow, J. R. (1987). The social and cultural implications of the VCR: How VCR use concentrates and diversifies viewing (Doctoral dissertation, University of Pennsylvania, 1987). *Dissertation Abstracts International, 48*, 03A.

Duff, R. W., & Hong, L. K. (1982). Quality and quantity of social interactions in the life satisfaction of older Americans. *Sociology and Social Research, 66*, 418-434.

Graney, M. J., & Zimmerman, R. M. (1980). Health self-report correlates among older people in natural random sample data. *Mid-American Review of Sociology, 5*, 47-59.

Greenberg, B. S., & Heeter, C. (1987). VCRs and young people. *American Behavioral Scientist, 30*, 509-521.

Gunter, B., & Levy, M. R. (1987). Social contexts of video use. *American Behavioral Scientist, 30*, 486-494.

Harvey, M. G., & Rothe, J. T. (1985-1986). Video cassette recorders: Their impact on viewers and advertisers. *Journal of Advertising Research, 25*(6), 19-27.

Havighurst, R. J., Neugarten, B. L., & Tobin, S. S. (1968). Disengagement and patterns of aging. In B. L. Neugarten (Ed.), *Middle age and aging: A reader in social psychology* (pp. 161-172). Chicago: University of Chicago Press.

Klippel, R. E., & Sweeney, T. W. (1974). The use of information sources by the aged consumer. *Gerontologist, 14*, 163-166.

Lambert, Z. V., & Durand, R. M. (1975). Some precautions in using canonical analysis. *Journal of Marketing Research, 12*, 468-475.

Lefcourt, H. M. (1982). *Locus of control: Current trends in theory and research* (2nd ed.). Hillsdale, NJ: Erlbaum.

Levenson, H. (1974). Activism and powerful others: Distinctions within the concept of internal-external control. *Journal of Personality Assessment, 38*, 377-383.

Levy, M. R. (1987). Some problems of VCR research. *American Behavioral Scientist, 30*, 461-470.

Maddox, G., & Eisdorfer, C. (1962). Some correlates of activity and morale among the elderly. *Social Forces, 40*, 254-260.

Mikawa, J. K., Nordin, K., & Eyman, J. (1986). The self-consciousness scale and locus of control. *Psychological Reports, 59*, 939-942.

Roe, K. (1987). Adolescents' video use. *American Behavioral Scientist, 30*, 522-532.

Rotter, J. B. (1954). *Social learning and clinical psychology*. Englewood Cliffs, NJ: Prentice-Hall.

Rotter, J. B. (1966). Generalized expectancies for internal versus external control of reinforcement. *Psychological Monographs, 80*(1), 1-28.

Rubin, A. M. (1983). Television uses and gratifications: The interactions of viewing patterns and motivations. *Journal of Broadcasting, 27*, 37-51.

Rubin, A. M. (1986). Television, aging and information seeking. *Language & Communication, 6*(1/2), 125-137.

Rubin, A. M., & Bantz, C. R. (1987). Utility of videocassette recorders. *American Behavioral Scientist, 30*, 471-485.

Rubin, A. M., & Bantz, C. R. (1988). Uses and gratifications of videocassette recorders. In J. Salvaggio & J. Bryant (Eds.), *Media use in the information age* (pp. 181-195). Hillsdale, NJ: Erlbaum.

Rubin, A. M., & Rubin, R. B. (1982). Contextual age and television use. *Human Communication Research, 8*, 228-244.

Rubin, A. M., & Rubin, R. B. (1985). Interface of personal and mediated communication: A research agenda. *Critical Studies in Mass Communication, 2*, 36-53.

Rubin, A. M., & Rubin, R. B. (1986). Contextual age as a life-position index. *International Journal of Aging and Human Development, 23*, 27-45.

Rubin, R. B., & Rubin, A. M. (1982). Contextual age and television use: Reexamining a life-position indicator. In M. Burgoon (Ed.), *Communication yearbook 6* (pp. 583-604). Beverly Hills, CA: Sage.

Rubin, R. B., Perse, E. M., & Barbato, C. A. (1988). Conceptualization and measurement of interpersonal communication motives. *Human Communication Research, 14*, 602-628.

Schoenbach, K., & Hackforth, J. (1987). Video in West German households. *American Behavioral Scientist, 30*, 533-543.

Schutz, W. C. (1966). *The interpersonal underworld*. Palo Alto, CA: Science and Behavior Books.

SPSS, Inc. (1988). *SPSS-X user's guide* (3rd ed.) Chicago: Author.

Tesch, S., Whitbourne, S. K., & Nehrke, M. F. (1981). Friendship, social interaction and subjective well-being of older men in an institutional setting. *International Journal of Aging and Human Development, 13*, 317-327.

Wenner, L. (1976). Functional analysis of TV viewing for older adults. *Journal of Broadcasting, 20*, 77-88.

Williams, F., Phillips, A. F., & Lum, P. (1985). Gratifications associated with new communication technologies. In K. E. Rosengren, L. A. Wenner, & P. Palmgreen (Eds.), *Media gratifications research: New perspectives* (pp. 241-252). Beverly Hills, CA: Sage.

7

Subjective Differences in the Use and Evaluation of the VCR

MILTON J. SHATZER
and
THOMAS R. LINDLOF

Since the late 1970s, a sizable and ever-growing body of research has focused on the videocassette recorder. With penetration in U.S. television households at approximately 60%, the VCR is rapidly becoming a permanent accessory to television viewing. Research has been focused on the VCR as an object of social and cultural definition, on its impact on social formation, and on its impact on individuals, groups, or societies (Levy, 1987a). Implicit in most investigations is the assumption that the majority of people use, perceive, understand, and evaluate the VCR in much the same way.

Although this approach is informative in describing the average VCR user (or normative use of the VCR in a general population) it offers little in understanding how individuals systematically differ in their *subjective* perceptions of this new technology. Therefore, the primary goal of the research reported here is to identify and characterize types of VCR users to improve our theoretical understanding of individual experience with the VCR. Our study was designed to uncover conceptual categories corresponding to the coherent ways that individuals perceive the utility and significance of VCR technology in their own circumstances.

Researchers in a number of social sciences have recognized that individuals cluster together in their subjective perception of phenomena according to certain *schemata* (Brown, 1980; Cattell, 1952;

Kerlinger, 1986; Stephenson, 1967). These schemata are the result of what Stephenson (1976) has termed *convergent subjectivity* on the part of particular groups within a larger population. McKeown and Thomas (1988, p. 12) define this subjectivity as "a person's communication of his or her point of view." They go on to say that subjectivity is anchored in self-reference, i.e., a person's frame of reference. Patterns of use and perceptions of the VCR are not homogeneous across the population, but vary in ways similar to individual political preferences, esthetic values, product preferences, and so on.

In this study, Q methodology was used to accomplish the goal of discovering the clusters or types of VCR users. Q methodology links both qualitative and quantitative methodologies (McKeown & Thomas, 1988). It is a powerful heuristic technique extremely useful in theory building because it uncovers the subjectivity operating within the individual.

STUDYING HOME VIDEO: FROM USAGE TO PERCEPTIONS

As with most household media innovations, the first wave of VCR inquiry focused at the descriptive level: who uses the VCR for what purposes, and for how long? In other words, the studies of roughly ten years ago attempted to quantitatively establish some baseline notions of what early adopters were doing with their machines. Given the infant state of the prerecorded video market at the time and the still-unsettled format standardization issue, it is not surprising that time-shifting behavior was a prime object of interest. For instance, it was found that network affiliate programming, especially prime time, was the most frequent source for off-air recording (Agostino & Zenaty, 1980; Levy, 1981; Levy, 1980b), that the VCR was used for replaying off-air recordings approximately 3.5 times a week (Levy, 1980a), and that the most regularly scheduled programs (like soap operas) were viewed soonest after taping, compared to special event programming (Levy, 1981; Levy & Fink, 1984). According to these and other data from the first studies, television viewing patterns were not substantially different after VCR adoption. At most, its introduction seemed to add on a few hours of weekly viewing and delay somewhat the viewing of content that the individual desired to view anyway.

Diffusion characteristics. As the diffusion of video recorders exploded through the mid-1980s, attention shifted to two foci: (1) classical innovation adoption research, and (2) examination of the use of technical features, including visual-scan fast-forward (*zipping*), recording pause (*zapping*, mostly during commercial breaks), and remote control channel selectors. The two are related, in that diffusion research is often directed at explaining the personal characteristics (e.g., demographic and life-style) of those individuals who adopt and apply an innovation. It was found that certain personal characteristics seemed to be good discriminators regarding which household members would be interested in and use the more advanced features of VCRs. Thus, VCR research began to develop some theory of its own by noticing, and then postulating, the sources of differences in adopter characteristics and behavior.

During this period, VCR-adopting households could be characterized as richer in most types of media ownership, including multiple television receivers, cable, and computers (Greenberg & Heeter, 1987; Gunter & Svennevig, 1987; Krugman, 1985). VCR owners were younger, more affluent, higher in education and occupational status, and more likely to live in urban areas (see Klopfenstein & Swanson, 1987). As summarized in a recent review (Lindlof, in press), VCR households reported a decline after adoption in movie-going and selected other out-of-home activities as well. The fact that households with young children are overrepresented among VCR owners may partly account for that pattern. There is not as clear an indication of a VCR effect among adolescents, who have stronger social motivations for out-of-home entertainment activity (Greenberg & Heeter, 1987; Hughes & Dobrow, 1988). Lastly, there is mixed evidence about whether gross VCR usage decreases with length of ownership, indicating a novelty fade effect. Prerecorded tape rentals may well decline shortly after the VCR is purchased, and then stabilize (Donohue & Henke, 1985). But off-air taping appears to increase with longer tenure of VCR ownership (Donohue & Henke, 1985; Klopfenstein & Swanson, 1987). It must be emphasized that these findings are few and inconclusive. Not only are larger and more diverse data sets needed, but closer inspection of such factors as skill learning and product perception would reveal more about the meaning of these findings and their long-term determinants.

Gender differences. Increasingly, gender has emerged as a characteristic that differentiates some aspects of VCR adoption and usage. Men are reported to have the most influence in adopting the VCR (Harvey & Rothe, 1985/1986). The male head of household was cited in one study as more likely to play tapes, although the female head of household was a close second in recording decisions (Donohue & Henke, 1985). Commercial avoidance by means of the act of zapping is more frequent for males (Donohue & Henke, 1985; Heeter & Greenberg, 1985; Klopfenstein & Swanson, 1987), but there seems to be no gender difference for zipping (Klopfenstein & Swanson, 1987).

Recently published qualitative studies provide more finely detailed portrayals of gender-based VCR usage in specific family situations. In an interview study of 18 British families, Morley found that the women usually avoided operating the video recorder and would defer to other family members, typically the fathers and sons, in matters of tape rentals, recording decisions, and playback occasions. This orientation, Morley claimed, was consistent with the gendered bifurcation of work and pleasure roles in the home. The husband's dominance over viewing and using the remote control, often free of competing responsibilities, relegated women to stealing their viewing opportunities at times when there would be no conflict with male selections.

Another qualitative study, this one relying more on in-home observations of family behavior (Lindlof, Shatzer, & Wilkinson, 1988), largely substantiates this pattern of female VCR usage. Not all of the women, however, were deficient in their ability to operate the VCR. Moreover, most of the women displayed what the authors called a *user competence.* Women learned as much about the VCR as they felt was needed to accomplish their personal and household goals. Despite occasional frustrations with the technology, the women's methods for accommodating video and television in their family routines were often inventive and highly practiced. It is clear that gender differences in VCR use is an emerging area of research that is running ahead of our ability to conceptualize the underlying mechanisms.

Children's VCR usage. Another area that has seen an increasing amount of attention, but also without much theoretical coherence, is that of children's interactions with VCR technology and its associated products. Ethnographic research suggests that children often form a special relationship with video, as compared to normal television reception. Their use of production and story conventions, their viewing attention, and their tolerance of multiple viewings of individual programs seem markedly different with the VCR (Lindlof & Shatzer, in press; Lindlof, Shatzer, & Wilkinson, 1988). Moreover, many of the young children show a high level of operational facility with VCR features. On the other hand, a survey study of the needs satisfied by seven types of consumer media for a sample of Israeli children did not find that VCRs have a special identity (Cohen, Levy, & Golden, 1988).

Recent studies have examined the viewing patterns and frequency of children and adolescents with VCR access, as well as the degree of parental control involved in that access. Thus far, there is no evidence that adolescents in VCR households spend more time viewing television than comparable non-VCR samples (Brown, Bauman, Lentz, & Koch, 1987; Greenberg & Heeter, 1987). White children seem to have greater access to VCRs than either black or Hispanic children (Blosser, 1986; Brown, et al., 1987; however, see Greenberg & Heeter, 1987). Parental employment and family structure may be more salient factors in influencing children's VCR access (Greenberg & Heeter, 1987; Lin & Atkin, 1988).

Addition of a VCR seems to alter the social context of children's viewing. Children view alone more often when playing back tapes than during off-air viewing occasions (Gunter & Svennevig, 1987). Regarding parental control, Kim, Baran, and Massey (1988) reported that parents and their children (6 to 16 years) are in substantial agreement about the lack of prescreening of tapes, the amount of conversation during viewing (no more with VCRs than without), the frequency of family viewing (no more with VCRs), perceptions of broadcast television control, and the use of tapes and televiewing as a reward for children's good behavior. However, parents perceived greater control in their use of the VCR, as compared to their children's perceptions. Lin and Atkin (1988) also found little difference between VCR and non-VCR homes in their reported parental mediation of and rule-making for adolescent viewing. They did find that ownership of new media (cable and VCR) was a negative predictor of parental mediation, including behaviors that prohibit or encourage certain kinds of programming. In general, it appears that parents exercise minimal behavioral and interpretive control over adolescent viewing of tapes (Greenberg & Heeter, 1987; Kim et al., 1988; Roe, 1987).

Activity in VCR usage. In the interest of basing VCR-user inquiries in a theoretical framework of human behavior, uses and gratifications concepts have lately been applied. These concepts are grounded in the active-audience precept, which assumes that individuals make media use decisions on a conscious, rational, and goal-directed basis. This orientation seems well-suited to the domain of VCR usage for two reasons: video users not only have a wider array of program alternatives from which to choose, they must also engage in active procedures for planning and engaging in the viewing of those alternatives.

Levy (1987b) sought to understand the orientations of VCR owners in terms of typology of audience activity. This typology posits a structure consisting of two dimensions, each having three values: (1) a qualitative dimension (the activity orientations of selectivity, involvement, and utility); and (2) a temporal dimension (locating the activity orientations before, during, or after exposure). He found that VCR owners display a relatively high degree of activeness for all nine measures (three qualitative values x three temporal values). Particularly strong endorsements were reported for the interest-arousing and memorable qualities of video content. Correlations between and within the temporal values were generally weak, albeit consistent. The author concluded that levels of activity may vary widely within individuals and across situations.

In another recent study of audience activity, Rubin and Bantz (1987) employed a canonical-correlation analysis to study the relationships between (1) selective and instrumental VCR uses, and (2) demographic characteristics and other media behaviors. This study did not test for degree or frequency of activeness, but did find differences in VCR user activity that systematically relate to aspects of the respondents' interpersonal communication and demographic traits. The nature of the canonical roots themselves seem suggestive of discrete life-styles in which the VCR may play differential roles.

Research questions. As revealed in this review, research of VCR usage has passed through several phases in its brief history. The initial studies simply set out to document the actual VCR activities of early users, especially as they involved time-shifting functions. Following shortly afterwards were studies focusing on such diffusion-related concerns as the demographic, time-displacement, and media usage characteristics of adopters. It should be noted that this category of inquiry is still underway even as VCR penetration approaches two-thirds of all American households.

The most recent studies have been motivated by both social concerns and theoretical considerations. Although there are others in the literature, we have identified research in the areas of gender-related differences, children's VCR interactions (including parental control), and the activity dimensions of VCR users. All three of those areas are marked by indications that significant differences of *perception* of the technology, its associated products, and its applications exist among the populations that have been studied. For example, studies that have considered gender point to male-female differences in overall orientation to home video. Similarly, parents and children vary in their perceptions of VCR control and decision making. While a child's age seems to matter in the degree of video usage control exerted within families, the age variable may well be a surrogate for more fundamental perceptions of adolescents' expanded social experience and behavioral options as they relate to VCR access. Lastly, research of VCR activity has focused on respondents' perceptions of their motives, the salience of which may depend on situational factors.

We would argue that it is important to begin investigating in a more direct fashion the organization of individuals' perceptions of home video. The intent of this line of inquiry is twofold: (1) How can we characterize the perceptual dimensions, and their constituent properties, that people use to evaluate VCR phenomena? (2) Do those perceptual dimensions correspond to or somehow transcend demographic or other characteristics and traits? This study therefore represents an inductive effort to identify the coherent perceptual dimensions that types of individuals use to evaluate their VCR engagement.

METHOD

Respondents. Fourteen two-parent families who owned VCRs were selected for this study from the greater Lexington, Kentucky area. Each family had at least one child 13 years of age or younger living at home. This purposive sample was selected so that individual experience with the VCR, perceptions of family use, and family communication patterns could be studied. Subjects were recruited through personal referrals. All families were either from middle or upper-middle class backgrounds and did not represent any minority groups. Average age of the participants was 35.5 (SD = 4.23). For 10 of the 14 families, the wives worked outside the home. Eleven of the families had two children, the remaining three had three children each. Average age of the children was 7.3 (SD = 4.30).

Data collection. Data were collected in the homes of the participating families. Collectively, the husband and wife completed a questionnaire that asked for information about the family, ownership and use of household media, information about the ownership of the VCR, and information about the ownership of a video camera (if applicable). In addition to completing the questionnaire, both the husband and wife were asked to individually complete two Q sorts. One sort dealt with individual perceptions of the VCR and the second with those of the family. Only the results from the individual sort are reported here.

The Q deck for the individual sort contained 90 statements (items) ranked on a continuum anchored by "most like me" and "least like me." The 28 participants were asked to sort the 90 items in an approximately normal distribution with a total of 11 columns (or categories). The extreme categories each contained three cards, adjacent columns four cards, with the entire sort having the following configuration: 3 4 7 10 12 14 12 10 7 4 3. Items differentiating the types or clusters of individuals are generally found in the two extreme categories (or columns) with more neutral items near the center.

In Q methodology the number of items sorted is the *Q-sample* ($N = 90$) and the number of participants correspond to the *variables*. That is, in Q-technique individuals are treated as variables and are factor analyzed (as opposed to the more familiar R factor analysis). Individual *types* are the resultant factors that emerge.

In this study, the Q-sample was quasi-naturalistic and quasi-structured according to McKeown and Thomas's (1988) terminology. That is, items were collected from sources other than the respondents and were designed to be representative of statements drawn from previous research. The statements represented findings from the wide variety of research previously cited in the literature review. The statement items exemplified the following seven categories: (a) VCR viewing practices, (b) organization and handling of VCR and related items, (c) VCR-related decision making and preferences, (d) knowledge and satisfaction regarding VCR operation, (e) general VCR attitudes, (f) general television attitudes, and (g) individual user styles. All items were pretested to eliminate redundant or unclear statements.

RESULTS

Procedures. The Q data were analyzed using the *QUANAL* (Van Tubergen, 1980) computer program. Items were scored with values from 1 to 11 (11 = "most like me," 1 = "least like me"). These rankings were correlated and then factor analyzed to discover the clusters of subjective perceptions among the respondents. Factors were extracted using principal axis analysis with squared multiple correlations on the diagonal. Eigenvalues greater than 1.00 were used as the minimum criterion for factor selection. Factors were then rotated using Varimax rotation to make them more interpretable. Although two factors were suggested by the scree test, three factors were selected as being theoretically meaningful (McKeown & Thomas, 1988; Norusis, 1985). All 28 respondents loaded on these three factors which explain 68% of the trace variance and 34% of total variance. The factors represent the groups whose sorting patterns of the items were similar.

The *QUANAL* program constructs the typical response pattern for each group or type (Van Tubergen, 1980). This composite response is expressed in z-scores. Conventionally, greatest interpretive weight is given to items which receive a z of plus or minus 1.0 in a type, i.e., those sorted in the most extreme two or three categories (Van Tubergen & Boyd, 1986). Differences between types are indicated by a typal z difference greater than 1.0.

The three types of VCR users that emerged were not given labels (as typically done in factor analysis) because the complexity of each type cannot be described adequately in one or two words. Clusters of individuals are thus identified as either Type 1, 2, or 3. Q items selected as most like the types (positive items) and least like the types (negative items) are reported in Tables 1, 2 and 3 respectively. For the following typal descriptions, corresponding Q item numbers are reported in brackets.

Type 1. Type 1s reported that they definitely like to do other activities while watching television and (to a lesser degree) the VCR [Q items 87 and 90]. They responded most positively to the item that they generally are involved in activities (such as reading or sewing) while they are watching television [87]. Further, they reported the most negative response to the item that they don't like to be doing something while watching television [75].

In addition, Type 1s reported using their VCR for child-focused activities. That is, they use the VCR a lot to record programs for the children [15]; and they enjoy watching a tape with their children even when there are no other adults around [54]. They (or their spouses) tape shows for their children that the children did not request [38]. In addition, they conveyed in a positive way that most of the programs they tape are for their children.

Type 1s reported that the VCR makes it easier for them to watch their favorite programs [67]. Similarly, they use the VCR a lot to record and playback their favorite programs (i.e., a decided time-shifting orientation) [19]. This group plans what to tape in upcoming programs by using a program guide [43].

Table 7.1 Z-scores for Type 1 Individuals (compared with Types 2 and 3)

		Types	
Item	1	2	3

Positive descriptors

Item	1	2	3
87. I'm generally involved in activities (such as reading, sewing, etc.) while watching TV.	**2.4**	−1.4	0.6
15. I use the VCR a lot to record programs for the children.	**1.9**	0.6	−0.5
67. The VCR makes it easier for me to watch my favorite programs.	**1.6**	0.4	−0.4
43. I plan what programs to tape by looking at the upcoming shows in a television guide (or similar publication).	**1.4**	−0.4	0.9
54. I enjoy watching a tape with my children even when there are no other adults around.	**1.3**	1.0	0.3
90. I'm generally involved in activities (such as reading, sewing, etc.) while watching tapes.	**1.2**	−1.3	−0.0
10. I don't like to watch the same recorded programs more than once.	**1.2**	−0.9	**1.1**
8. I often watch the same tapes my children like to watch.	**1.2**	**2.0**	−1.0
82. The VCR is a lot more valuable if one has cable television.	**1.2**	−0.1	**1.4**
71. I feel I have more control over the television now that we have the VCR.	1.1	1.1	0.7
51. I like the freedom the VCR allows to set my own schedule.	**1.1**	**1.8**	0.7
19. I use the VCR a lot to record and playback my favorite regular programs.	**1.0**	−0.2	−0.5
38. My wife/husband or I will tape shows for the children that they have not requested.	**1.0**	0.6	0.2
31. The VCR is a lot more valuable if one has a remote control.	**1.0**	0.9	0.7

Negative descriptors

Item	1	2	3
25. Occasionally, I rent adult tapes.	**−2.7**	**−1.2**	−1.0
36. I like to rent and watch unconventional or "offbeat" tapes.	**−2.1**	−0.6	0.8
20. I use the VCR a lot to record daytime soap operas.	**−1.5**	**−1.9**	**−2.9**
52. I use my VCR to rewatch programs and view them critically.	**−1.5**	−0.3	−0.1
16. I like to watch science-fiction movies, programs, etc.	−1.4	**1.1**	**1.2**
75. I don't like to be doing something else while watching television.	**−1.4**	1.7	0.2
78. I like to own new electronic devices when they come on the market.	**−1.4**	**−1.2**	0.1
12. I like to record important news events that may have historical significance to have in my VCR library.	**−1.1**	−0.7	−0.2

Type 1 individuals definitely do not rent adult tapes (even though all types responded negatively to this item, Type 1s are much more negative than the other two types) [25]. They do not like to rent and watch unconventional or offbeat tapes [36]; and they decidedly do not like to use the VCR to rewatch programs to view them critically [52]. Further, they do not like to watch science-fiction movies or programs [16]. They do not record important news events that may have historical significance for their videotape library [12].

The Type 1 cluster consisted of 10 individuals (6 females and 4 males). Females reported their occupations as: homemaker or housewife (4), teacher, and nurse; and males as: engineer (2), television station manager, and university administrator.

Type 2. For Type 2s, price was reported as an important consideration in the purchase of the VCR [64]. Moreover, prepurchase information came from people who owned a VCR rather than from articles in newspapers and magazines [65, 66].

Type 2 was most positive in reporting they watch tapes their children like to watch (being similar to Type 1 in this respect) [8]. This group was most positive in reporting that they liked the freedom the VCR allows to set one's schedule (again, similar to Type 1) [51]. These individuals definitely do not like to be doing something else while watching television [75]. They have a favorite spot (chair, sofa, etc.) from which they like to view [88]. They decidedly do not like to get up and manually operate the VCR while playing tapes [29]. For this type, watching movies on a VCR is as good or better than a movie theater [70]. Further, they like the privacy of the VCR [69].

Like Type 3s, Type 2s like to watch science-fiction movies and other programs, [16]; they are not careful in organizing their tapes [28]; nor do they keep tapes of their favorite time-shifted programs in a video library [27]. This group definitely does not use the VCR a lot to record recent movies for friends [17].

Like Type 1s, Type 2s reported they do not like to own new electronic devices when they come on the market [78]. They markedly are not able to operate every feature on their VCR [56]. This group definitely does not know as much about their VCR as they would like [55]. Finally, they definitely do not believe that men have a better grasp on how the VCR functions [60].

The Type 2 cluster consisted of 10 individuals (5 females and 5 males). Females reported being employed as: humanities professor, clerk, real estate agent, nurse, and office worker; and the males as:

Table 7.2 Z-scores for Type 2 Individuals (Compared with Types 1 and 3)

Item	Types 1	2	3
Positive descriptors			
64. Price was an important consideration in the purchase of our VCR.	0.4	**2.3**	**1.3**
8. I often watch the same tapes my children like to watch.	**1.2**	**2.0**	−1.0
51. I like the freedom the VCR allows to set my own schedule.	**1.1**	**1.8**	0.7
75. I don't like to be doing something else while watching television.	−1.4	**1.7**	0.2
65. Most of the information I got about the VCR before its purchase came from people who owned one.	0.6	**1.6.**	−0.1
88. I have a favorite spot (chair, sofa, etc.) from which I like to view.	0.8	**1.3**	−0.3
29. I don't like to get up and manually operate the VCR while playing back tapes.	0.3	**1.2**	0.9
71. I feel I have more control over the television now that we have the VCR.	**1.1**	**1.1**	0.7
70. For me, watching movies on a VCR is as good or better than in a movie theater.	−0.9	**1.1**	0.9
16. I like to watch science-fiction movies, programs, etc.	−1.4	**1.1**	**1.2**
69. I like the privacy of the VCR.	−0.6	**1.0**	0.6
54. I enjoy watching a tape with my children even when there are no other adults around.	**1.3**	**1.0**	0.3
Negative descriptors			
20. I use the VCR a lot to record daytime soap operas.	**−1.5**	**−1.9**	**−2.9**
21. I use the VCR a lot to record sports programs.	**−1.4**	**−1.6**	−0.6
28. I am careful about organizing our tapes.	−0.1	**−1.4**	**−1.7**
87. I'm generally involved in activities (like reading, sewing, etc.) while watching TV.	2.4	**−1.4**	0.6
17. I use the VCR a lot to record recent movies for my friends.	−0.2	**−1.3**	−0.7
90. I'm generally involved in activities (like reading, sewing, etc.) while watching tapes.	1.2	**−1.3**	−0.0
78. I like to own new electronic devices when they come on the market.	**−1.4**	**−1.2**	0.1
56. I am able to operate every feature on my VCR.	−0.5	**−1.2**	1.4
25. Occasionally, I rent adult tapes.	**−2.7**	**−1.2**	**−1.0**
60. I believe men have a better grasp as to how the VCR functions.	−0.4	**−1.2**	−0.2
27. I keep tapes of favorite time-shifted programs in a video library.	−0.2	**−1.1**	**−1.1**
55. I know as much as I want to know about my VCR.	0.3	**−1.1**	0.3
66. Most of the information I got about the VCR before its purchase came from articles in newspapers and magazines.	−0.5	**−1.1**	0.9

computer programmer, service manager, music professor, and repairman/electrician.

Type 3. Type 3s described themselves as using the VCR a lot to view rental tapes [14]. They also conveyed that they always view the programs they have taped [3]. They like to zap out commercials when they are taping from programs that they are also watching [2]. They flip channels so they can watch more than one program at a time [86]. This group likes to view their favorite tapes more than once [4]. However, they do not like to watch the same recorded programs more than once [10].

This type reported being able to operate every feature on their VCR [56]. They conveyed that if they like to own something, they generally go out and buy it [79]. When they buy a new electronic device, they like it to come fully equipped [80]. They unequivocally do not feel intimidated by new electronic devices [83], and for them the VCR was not difficult to learn to use [57]. Like Type 1s, Type 3s reported that the VCR is a lot more valuable if one has cable television [82].

Type 3s do not preview all their children's tapes before they allow them to be viewed. They did not identify with the statement that renting a movie on tape allows them to study the movie in detail [53]. Although all types represented in this sample do not use the VCR a lot to record daytime soap operas, Type 3s had the most negative response to this statement [20]. Like Type 2s, Type 3s described themselves as not careful about organizing tapes [28], and they do not keep tapes of favorite time-shifted programs in a video library [27].

The Type 3 cluster consisted of 8 individuals (3 females and 5 males). Females reported being employed as: laboratory clerk, medical records clerk, and doctor's office worker, and males as: engineer, professor, pharmacist, sales representative, and store manager.

Consensus among all three types. Aside from the individual typal differences, there was consensus from all three types on a number of items. All types rent tapes of movies that they did not get a chance to see in the local theaters, like to zip through commercials when watching time-shifted tapes, like to rent tapes from stores near their homes, and feel that they have more control over the television now that they have the VCR.

On the negative side, all three types do not use the VCR a lot to record music videos, nor do they use the VCR because they can see and hear their favorite songs at any time. Further, all three types do not believe the VCR to be the responsibility of the woman of the house, nor do they rent or buy a tape for instructions for jobs done in the house or outdoors. Moreover, they do not use the VCR a lot to record classic old

Table 7.3 Z-scores for Type 3 Individuals (compared with Types 1 and 2)

	Types		
Item	1	2	3
Positive descriptors			
14. I use the VCR a lot to view rental tapes.	0.8	0.7	**2.4**
3. I always view the programs that I have taped.	0.4	0.5	**1.5**
82. The VCR is a lot more valuable if one has cable television.	1.2	−0.1	**1.4**
56. I am able to operate every feature on my VCR.	−0.5	−1.2	**1.4**
64. Price was an important consideration in the purchase of our VCR.	0.4	2.3	**1.3**
16. I like to watch science-fiction movies, programs, etc.	−1.4	1.1	**1.2**
4. I like to view my favorite tapes more than once.	−0.8	0.4	**1.2**
2. I like to zap out the commercials when I am taping a program that I am also watching.	−0.1	0.4	**1.1**
79. If I would like to own something, I generally go out and buy it.	0.1	−0.5	**1.1**
80. When I buy a new electronic device, I like it to come fully equipped.	−0.7	0.1	**1.1**
86. I flip channels so I can watch more than one program at a time.	−0.8	−0.9	**1.1**
10. I don't like to watch the same recorded programs more than once.	1.2	−0.9	**1.1**
Negative descriptors			
20. I use the VCR a lot to record daytime soap operas.	**−1.5**	**−1.9**	**−2.9**
83. I sometimes feel intimidated by the new electronic devices that are currently on the market.	−0.1	0.0	**−2.1**
57. For me, the VCR was difficult to learn to use.	−0.1	−0.3	**−1.9**
28. I am careful about organizing our tapes.	−0.1	**−1.4**	**−1.7**
39. I preview all of the children's tapes before I allow them to view them.	0.0	0.7	**−1.4**
27. I keep tapes of favorite time-shifted programs in a video library.	−0.2	**−1.1**	**−1.1**
25. Occasionally, I rent adult tapes.	**−2.7**	**−1.2**	**−1.0**
53. Renting a movie on tape allows me to study the movie in detail.	−0.7	0.3	**−1.0**

films, nor do they rent or buy a tape for self-improvement, such as an exercise tape.

All of the positive consensus items centered on convenience (i.e., the saving of time and effort). Negative items, on the other hand, demonstrated a disregard for specialized content (even content that would fit within their life-style, e.g., instructional and how-to tapes). Further, all three types conveyed that they do not believe the VCR to be the responsibility of the woman of the house. However, it is not known whether the VCR should be within the male's domain or whether a more egalitarian sharing of VCR tasks is perceived.

DISCUSSION

In answer to the first research question, three major perceptual dimensions used to evaluate the VCR phenomenon were uncovered. Like other phenomena, VCR use varies according to the subjective perceptions, or operant subjectivity, of the users. In answer to the second question, these dimensions appear to transcend traditional demographic characteristics or traits and are more attuned to general lifestyle orientations.

Type 1 individuals are definitely more *polychronic* (following Hall, 1959) in their viewing of television and (to a lesser extent) the VCR. That is, they are generally involved in other activities while viewing. Type 1s are very interested in using the VCR as both an instrument of convenience and for structuring children's viewing. This type is more of a time-shifter and plans what to tape by using a television guide. Much of their VCR use is child centered—they tape a great deal for their children and watch tapes with their children. This type is also more conventional in their tape rental and viewing behavior. They definitely do not rent adult tapes or offbeat tapes. Further, they do not use the VCR to rewatch and view tapes critically, they do not watch their favorite tapes more than once, they do not like science-fiction fare, nor do they record historical events for their VCR library.

Type 2 individuals are more price-conscious VCR consumers. They gather prepurchase information from people who own VCRs (and not from the mass media). These people are *monochronic* (Hall, 1959) viewers—they definitely do not like to be doing other things while watching television or watching tapes. They have a definite comfort orientation, i.e., they have a favorite viewing location and do not like to have to manually operate the VCR while replaying tapes. Moreover, they are more egocentric in their use of the VCR. They like the privacy and schedule-setting capabilities of the VCR. For them, watching movies on the VCR is as good or better than attending a movie theater.

Unlike the other two types, they watch the same recorded programs more than once. Type 2s watch the same tapes their children like to watch, and they often view with their children when no other adults are around. They do not tape a lot of movies for friends. In addition, this type does not know how to operate every feature on their VCR, but they would like to know more.

Type 3 individuals are avid tape renters. They conscientiously view the programs they have taped and view their favorite tapes more than once. Type 3s have a definite *technical* orientation. That is, they can operate every feature on the VCR, prefer to buy electronic devices fully equipped, are not intimidated by new electronic devices, like to play around with electronic gadgets, and for them the VCR was not difficult to learn to use. These people are zappers of commercials and they graze over the dial (i.e., flip channels). They definitely do not tape soap operas and seldom preview their children's tapes. Of the three groups, Type 3s are more prone to rent and watch unconventional or offbeat tapes. Most of their prepurchase information concerning the VCR comes from the mass media.

All three types share some perceptions in common. All like to rent tapes of films they missed in the theaters, zip through commercials in time-shifted tapes, like to rent tapes from outlets near their homes, and feel they have control of the television with the VCR. The types identified in this study do not record music videos; do not, in general, rent or buy instructional tapes; nor do they believe the VCR to be the responsibility of the woman of the house. These findings, however, may be a function of respondents used in this study. Greater generalizability awaits further empirical research.

Comparison with prior research. These typal characteristics illuminate some of the findings of our earlier ethnographic research (Lindlof et al., 1988; Lindlof & Shatzer, in press), and suggest underlying differences in individual perceptions of the VCR. In that study, most family members were observed to *satisfice* their VCR usage. That is, they learned just enough about their VCRs to accomplish desired tasks, but not enough to maximize the usage potential implicit in the machines' full array of features. This satisficing orientation was manifested in different ways by the individuals depending on their family roles and associated activity patterns.

For example, certain family members (mostly, but not exclusively females) could play back tapes and time-shift programs, but were not as adept at some of the more complex VCR operations. They were mostly interested in the convenience of the VCR for multiplying viewing opportunities in the midst of busy personal schedules. Although these individuals were often dependent on other family members for performing certain VCR operations, this dependency was seldom expressed as a problem. Even those family members who used more VCR features and were more proficient in that usage did not take advantage of all that they knew about the machine's features and functions.

This study's Type 1 has some characteristics consistent with this satisficing mode in the sense that the VCR is perceived as an effective way to view while involved in other activities. Moreover, Type 1 individuals domesticate the occasionally undesirable elements of video and broadcast content by watching with their children and by engaging in much advance planning of their taping. Thus, they appear to be homebodies whose world centers around hearth (including the television and VCR in contemporary U. S. society) and home. Type 2 individuals also integrate the VCR with family life where children are involved, but for them television and video represent experiences that are best enjoyed without competition from other tasks. They are also uncertain about their own competence as VCR operators.

Of the three types, the Type 3 individuals show the poorest fit to the satisficing profile. They are comfortable with all manner of electronic devices and rate usage of the most prominently interactive features of the VCR very highly. The Type 3 person appears more individualistic in the pursuit of gratifications through the control of technological options and program content.

In summary, these types reveal structures of meaning that may help explain why one person seeks a different, perhaps more efficient level of satisficing than another person. But we still do not know what it means to *maximize* the VCR's potential simply because we do not know what the optimal levels of knowledge or usage frequency in everyday practice might be. Type 3 persons *might* maximize more than they satisfice. On the other hand, there could well be more variance in the complexity and flexibility of their subjective schemas for home video than this study suggests, to say nothing of the variation in Type 3s actual usage patterns.

Like Rubin and Bantz (1987), our study provides further evidence of individual differences in media consumers with respect to the VCR. However, in opposition to Rubin and Bantz (1987), the typal dimensions revealed are *not* due primarily to demographic differences such as age or gender. The participants in this research are, for the most part, demographically homogeneous. That is, all participants are primarily middle-aged, white, middle to upper-middle class, and parents in a two-parent household. Yet three distinct types of VCR users emerged. The demographic differences that are present (i.e., gender and occupation) did not systematically affect cluster formation across all three types. Quite possibly the typal differences may be due to a general overall *life-style* orientation. This is suggested by the fact that 8 of the 14 families had both husbands and wives in the same cluster, thus sharing the same (or similar) orientation.

One interesting aside to this research is that different types of individuals have different sources of prepurchase information about the VCR. Traditional diffusion research (see Rogers, 1983) has looked at the innovation-decision process as undifferentiated in terms of individual involvement. This research suggests that there may be individual, subjective differences in this process as well.

In conclusion, this study provides a unique and insightful perspective to VCR research. It goes beyond traditional normative research that characterizes average VCR use and has identified at least three types of VCR users. The fact that individuals cluster together (based on similar perceptions) does not preclude participation in behaviors characteristics of other types. Rather, it illustrates types of individuals who share subjective perceptions of the VCR phenomenon. People's use of the VCR reflects personal, or possibly, greater life-style orientations. Future research needs to identify these orientations for a more comprehensive understanding of the VCR experience. The use of Q methodology, combined with a greater understanding of people's attitudes and values, is one of the ways in which this can be accomplished.

REFERENCES

Agostino, D., & Zenaty, J. (1980). *Home VCR owners' use of television and public television: Viewing, recording, and playback.* Washington, DC: Office of Communication Research, Corporation for Public Broadcasting.

Blosser, B. J. (1986, May). *Ethnic differences in media use: A study of children, television, and other media.* Paper presented at the meeting of the International Communication Association. Chicago.

Brown, J. D., Bauman, K. D., Lentz, G. M., & Koch, G. G. (1987, May). *Young adolescents' use of radio and television in the 1980s.* Paper presented at the meeting of the International Communication Association. Montreal, Canada.

Brown, S. R. (1980). *Political subjectivity: Applications of Q methodology in political science.* New Haven, CT: Yale University Press.

Cattell, R. B. (1952). The three basic factor-analytic research designs—their interrelations and derivatives. *Psychological Bulletin, 49,* 499-520.

Cohen, A. A., Levy, M. R., & Golden, K. (1988). Children's uses and gratifications of home VCRs: Evolution or revolution. *Communication Research, 15,* 772-780.

Donohue, T. R. & Henke, L. L. (1985). *The impact of video cassette recorders on traditional television and cable viewing habits and preferences.* Washington, DC: National Association of Broadcasters.

Greenberg, B. S. & Heeter, C. (1987). VCRs and young people. *American Behavioral Scientist, 30,* 509-521.

Gunter, B. & Levy, M. R. (1987). Social contexts of video use. *American Behavioral Scientist, 30,* 486-494.

Gunter, B., & Svennevig, M. (1987). *Behind and in front of the screen: Television's involvement with family life.* London: John Libbey.

Hall, E. T. (1959). *The silent language.* New York: Doubleday.

Harvey, M. G. & Rothe, J. T. (Dec. 1985/Jan. 1986). Video cassette recorders: Their impact on viewers and advertisers. *Journal of Advertising Research, 25*(6), 19-27.

Heeter, C. & Greenberg, B. (1985). Cable and program choice. In D. Zillman & J. Bryant (Eds.), *Selective exposure to communication* (pp. 203-224). Hillsdale, NJ: Erlbaum.

Hughes, C. E. & Dobrow, J. R. (1988, May). *The VCR and the adolescent: Patterns of use.* Paper presented at the meeting of the International Communication Association, New Orleans.

Kerlinger, F. N. (1986). *The foundations of behavioral research* (3rd ed.). New York: CBS College Publishing.

Kim, W. Y., Baran, S. J., & Massey, K. (1988). Impact of the VCR on control of television viewing. *Journal of Broadcasting & Electronic Media, 32*(3), 351-357.

Klopfenstein, B. C. & Swanson, D. A. (1987, May). *An analysis of VCR adopter characteristics and behavior.* Paper presented at the meeting of the International Communication Association, Montreal.

Krugman, D. M. (1985). Evaluating the audiences of the new media. *Journal of Advertising, 14*(4), 21-27.

Levy, M. R. (1987a). Some problems of VCR research. *American Behavioral Scientist, 30,* 461-470.

Levy, M. R. (1987b). VCR use and the concept of audience activity. *Communication Quarterly, 35*(3), 267-275.

Levy, M. (1981). Home video recorders and time shifting. *Journalism Quarterly, 58,* 401-405.

Levy, M. R. (1980a). Home video recorders: A user survey. *Journal of Communication, 30*(4), 23 27.

Levy, M. (1980b). Program playback preferences in VCR households. *Journal of Broadcasting, 24*(3), 327-336.

Levy, M. R. & Fink, E. L. (1984). Home video recorders and the transience of television broadcasts. *Journal of Communication, 34*(2), 56-71.

Lin, C. A. & Atkin, D. J. (1988, May). *Parental mediation and adolescent uses of television and VCRs.* Paper presented at the meeting of the International Communication Association, New Orleans.

Lindlof, T. R. (in press). New communications media and the family: Practices, functions, and effects. In B. Dervin (Ed.), *Progress in communication sciences* (Vol. 10). Norwood, NJ: Ablex.

Lindlof, T. R. & Shatzer, M. J. (in press). VCR usage in the American family. In J. Bryant (Ed.), *Television and the American family.* Hillsdale, NJ: Erlbaum.

Lindlof, T. R., Shatzer, M. J., & Wilkinson, D. (1988). Accommodation of video and television in the American family. In J. Lull (Ed.), *World families watch television*. Newbury Park, CA: Sage.

McKeown, B., & Thomas, D. (1988). *Q methodology*. Sage University Paper series on Quantitative Applications in the Social Sciences, series no. 07-066. Beverly Hills, CA: Sage.

Morley, D. (1986). *Family television: Cultural power and domestic leisure*. London: Comedia Publishing Group.

Norusis, M. J. (1985). *SPSS^x advanced statistics guide*. Chicago: SPSS Inc.

Roe, K. (1987). *Adolescents' VCR use: How and why*. Paper presented at the meeting of the International Communication Association, Montreal.

Rogers, E. M. (1983). *Diffusion of innovations* (3rd ed.). New York: Free Press.

Rubin, A. M. & Bantz, C. R. (1987). Utility of videocassette recorders. *American Behavioral Scientist, 30*, 471-485.

Stephenson, W. (1967). *The play theory of mass communication*. Chicago: University of Chicago Press.

Van Tubergen, G. N. (1980). *QUANAL user's guide* (2nd ed., rev.). Lexington, KY: N. Van Tubergen.

Van Tubergen, G. N., & Boyd, D. A. (1986). Third-world images of U.S.: Media use by Jordanians. *Journalism Quarterly, 63*(3), 607-611.

Part III
The VCR and the Individual

8

Big Eyes But Clumsy Fingers: Knowing About and Using Technological Features of Home VCRs

AKIBA A. COHEN and LAURA COHEN

Home videocassette recorders (VCRs) have been around now for some time. In fact, since the late 1970s there has been an increasing penetration of the VCRs into households in many nations around the world, in both limited- and rich-media environments (Boyd & Straubhaar, 1985; Lin, 1987).

It has often been argued that the advent of the VCR age has brought with it a technological revolution in terms of how television, its predecessor, could be used (see, for example, Cubbitt, 1986; Gubern, 1985). In a purely theoretical sense this is clearly a strong argument concerning technological development. After all, ever since television became a standard household appliance in the late 1940s and early 1950s, there was not much that the viewer could do with the set other than watch a particular program, even if the choice of which program to watch has increased somewhat over the years.

Program quantity aside, television sets always had a channel selector, a knob (or two) that could adjust the brightness of the picture (and its hue or tint when color was introduced) and a volume control. Nothing more, even if television screens got larger (and also smaller—even miniature), television more portable, slimmer as well as more elegant, and had better sound quality (including stereo). Even a remote control unit does not allow for more features; it simply allows the viewer to operate the set from the sofa or bed. In short, a television is a television is a television.

Ordinarily, then, when purchasing a television set, whether in 1950 or today, almost all the consumer needed to consider was the brand name, which presumably stood for possible differential quality and its correlated price tag. The features of all television sets were essentially few and much the same at any given point in time. In other words, there was and still remains little variability within the product called *television*. And more importantly, perhaps, television was and still is an easy thing to use.

By contrast, the brief history of VCRs presents a totally different story (Schafer-Gross, 1986). In the dawn of the VCR era, in the mid-1970s, there were already different technologies to contend with. Early on, one had to choose between Betamax, developed by the Sony Corporation, and VHS, developed by JVC (both were developed at about the same time after the U-Matic 3/4-inch system, which was not really intended for home use).

Also, because of different television broadcast standards used around the world (for example, NTSC in the United States, PAL in most of Europe, and SECAM in France and the Soviet Union), it becomes necessary for the consumer to decide whether he or she wishes to be limited to viewing prerecorded cassettes from only one part of the world or to purchase a VCR that can play back material recorded in the other standards as well. It should be pointed out that this relatively costly option is rarely available to the U.S. consumer except in the major cities, probably for lack of interest by Americans in non-American programs and films, whereas such machines are more readily available in other parts of the world.

Moreover, during a very short period of time the variety of features in VCR technology has rapidly expanded. Thus, for example, while the early models enabled the recording of only one off-air program from a single channel, today's typical models make it possible to set up the machine to record several different programs, from different channels, during a period of two weeks or longer. This expanded capacity was often limited by the duration of recording time on a single cassette, which has now gradually increased to six hours on most machines, in the slowest of three speeds. And for those requiring even more storage space, a device is now available which automatically changes as many as six cassettes.

Additional features are also available. Thus, for example, virtually all VCRs now enable the user to run through the cassette at a fast speed (both forward and backwards)—known as picture searching or picture scanning—while the image remains fairly stable on the screen. Also, today's VCR can freeze or hold the picture to permit the viewer to look at details in the frame. Finally, some VCRs allow for slow motion, that is, the ability to see the picture at a less-than-normal pace. Thus, to paraphrase what was said earlier about television, a VCR is not necessarily a VCR is not necessarily a VCR. It seems that there is virtually no limit to what engineers can come up with, except perhaps to have the VCR record two programs simultaneously on the same cassette, but only time will tell if this too will be possible.

From a social rather than a technological perspective, the interesting question is not what is available for the VCR consumer, but rather what the consumer knows about the various possibilities that are available, what he or she succeeds in learning to use, and finally what the VCR owner actually utilizes.

In a recent paper summarizing issues in VCR research, Levy (1987) raises this point by suggesting research on "whether VCR owners use the full range of behaviors that video makes available" (p. 465). This chapter makes a modest attempt to focus on some of those issues.

What we would like to know is how much the typical consumer knows about the technological features of his/her machine, features which no doubt increase the cost paid for the new toy, (which in some countries can amount to a hefty expense). Since it is assumed that many of the features are hardly ever used, at least by many VCR owners, we would like to know if there are any dramatic changes in their use over time, just as with most other new toys.

Given the fact that the VCR is a moderately sophisticated electronic machine, which is present in many homes, often irrespective of socioeconomic factors, three additional foci were added to the study, when appropriate: the distinction between males and females; the difference among younger and older persons; and the possible contrast among people who originate from more- and less-technologically oriented societies.

It was expected that males, younger people, and those originating from more-technologically oriented societies would be more likely to be interested in, and hence knowledgeable about the technological features of the VCR, which in turn would lead them to use them more frequently. Men, it is suggested, would be more interested in VCR features than women, given the generally prevailing social bias suggesting that men are more technologically inclined to dealing with and repairing mechanical and electronic equipment. Younger people would be more inclined to appreciate the VCR features, once again, given the overall adaptability of young people to new technologies, compared with older folks. And people originating from cultures where high technology is more prevalent would be more inclined to be interested in it and less reluctant to deal with it than people whose main socialization was in environments that lack such technologies.

It is noteworthy that during the past decade, which has witnessed growing research on VCR use, almost no attention has been paid to people's knowledge and use of the technology or the social and cultural context of that practical knowledge. One exception to that scientific blind spot is a study by Lindlof, Shatzer and Wilkinson (1987), which examined how six families used their VCRs. Lindlof et al. report that many of the features of the machine are not used and that "VCR usage had actually regressed from frequent and diverse applications soon after purchase . . . to a simpler and more predictable regimen" (p. 8).

In terms of whether the VCR heralds a communication revolution of sorts, we could contend that if many VCR owners do not know about nor bother to use the available technological tricks the VCR has to offer, this would provide circumstantial evidence for the countervailing point of view that the VCR did not really create a revolution but rather, at most, should be considered an extension of television (for related arguments, see Cohen, Levy & Golden, 1988; Levy & Gunter, 1988).

The main objective of this chapter, then, is to report on two exploratory studies dealing with the degree of knowledge that VCR owners have about their video equipment and the extent to which they use some of its basic features.

THE TWO STUDIES

Both projects were conducted in Israel during 1987-1988. The first dealt with adolescents and included information on their own knowledge and use of VCR features as well as their perceptions of their parents' knowledge and use. Following that initial study, some questions of interest concerning parental knowledge and use of video technology remained. Since it was impossible, however, to establish contact with the parents of the original adolescent respondents, a second study was conducted in which a sample of adults, similar in demographic terms to the first sample, was interviewed by telephone. Obviously this would be less than ideal, but previous studies (e.g.,

Jennings and Niemi, 1974) have found that adolescents are often mature enough to be aware of and to be able to report on their parents' behaviors concerning a topic with which the adolescents themselves are also quite familiar.

The penetration rate of VCRs in Israel is at about the 40% level (Cohen, 1987). VCRs have been available in the country since the late 1970s. There is a wide array of models available, European, Japanese, and American made, with differing levels of technological sophistication (although it is virtually impossible to find today a stripped model with only a few features). The price paid for a VCR runs from about $1200 to more than $2000, including all the taxes (it should be noted that the average income for an Israeli family is about $1000 per month, before taxes). The demographics of VCR ownership indicate that there are no substantial differences by traditional socioeconomic variables, with only the higher educated owning slightly fewer sets. Moreover, VCR ownership is fairly evenly distributed throughout Israel. The main use made by VCRs in Israel is to view rented movies from over 300 rental outlets. Time shifting is done with much less frequency than in the United States, for example, mainly because of Israeli's limited television channel environment.

The Sample of Adolescents

Some 182 ninth grade students (ages 14-15) were interviewed by means of self-administered questionnaires in one high school in Jerusalem in May of 1987. The high school selected is composed of students from a cross section of the city's population, mostly middle-class and lower-class families. The ninth grade was chosen since children at that level have not yet chosen either an academic or vocational track, thus ensuring a relatively heterogeneous sample. The data reported below are from the 83 respondents (46% of the sample) who reported that their family owned a VCR.

The Adult Sample

Adult respondents were interviewed by telephone in June, 1988. Most of the 405 respondents were from Jerusalem but about 20% were from other towns in Israel. Forty graduate students in communication from Hebrew University conducted the interviews in their city of residence. Respondents were randomly selected from the telephone directories of the relevant cities. Each student-interviewer was assigned two pages of the telephone directories, using a skip interval of twenty pages. The numbers dialed used a skip interval of 12 numbers. Unlisted numbers were not used but since very few numbers are unlisted in Israel this fact would not cause a serious sampling problem. More than 90% of the interviews were completed. The data reported below are based on 159 respondents (39% of the sample) who reported owning a VCR.

Questionnaire Items

Both surveys asked about general media use, and included a filter question to determine whether the respondent's family owned a VCR. VCR owners were then questioned about their VCR knowledge and use.

In the sample of adolescents the questions began with the knowledge of the format of the VCR that the family owns. This refers to the VHS or Betamax alternatives. It should be noted that in Israel these two formats have been referred to for a long time as large and small cassettes, respectively, given the difference in their appearance. Hence the question was worded using *large* and *small*. The respondents were also asked whether their machine operates on the PAL, SECAM, or NTSC standard. For both questions the "don't know" option was available and read to the respondents.

Next the respondents were asked for their subjective assessment of how familiar they were with the functions of the various buttons on the VCR (all, most, some, or none), whether they know how to program the VCR to record off-air programs; the extent to which they personally use the picture search and pause function, and whether they watch or re-view parts of recordings over again.

The adolescents were also asked about their perception of their parents' (mother and father separately) knowledge and use of the VCR, that is, the children's beliefs about their parents' knowledge of the function of the various buttons, whether their parents can operate the machine by themselves, whether they can record programs off-air, and whether their parents are "scared" to use the VCR by themselves.

The demographic variables recorded were parents' ethnic origin (born in Israel, in Europe or the U.S., or born in another Middle Eastern/North African country), and the amount of time that the family owned the VCR (less than 1 year, 1-3 years, or more than 3 years).

Table 8.1 Adolescents' Responses by Gender, Fathers' Ethnic Origin and Time Owning a VCR (in percentages) [n=83]

	Total	Gender		Father's Ethnic Origin			Time Owning VCR		
		Boys	Girls	Israel	West	East	–1yr	1-3yrs	3+yrs
Knowledge									
Know size of cassettes	92.8	97.7	86.5	97.1	100.0	80.0	94.1	100.0	94.6
Know type of video system	31.3	47.5	10.8	36.4	35.0	26.1	23.5	30.0	36.1
Know function of *all* buttons[a]	78.3	90.7	62.2	79.4	75.0	80.0	70.6	81.0	81.1
Know to record off-air	87.8	92.9	81.1	91.2	80.0	87.5	82.4	95.0	91.9
Use of features									
Use picture search	61.4	67.4	54.0	52.9	70.0	64.0	70.6	61.9	56.7
Re-view program parts	76.8	76.2	75.7	70.6	80.0	80.0	87.5	66.7	75.7
Use pause function	51.3	51.2	50.0	42.4	52.6	62.5	50.1	55.0	47.2

a. The data reported in the table pertain to the *all* category only while the values of X^2 reported in the text pertains to the complete spread of the variable.

In the sample of adults, after determining VCR ownership, the respondents were asked about the VCR format, using both the size and trade name (VHS versus Betamax), the type of system (PAL/SECAM/NTSC), the knowledge of the function of the various buttons, and the extent to which they actually use the various functions (all, most, few, only on/off function, none).

The demograhics in the sample of adults included gender, age, educational level (incomplete high school, complete high school, post high school), ethnic origin, and amount of time the VCR was owned.

FINDINGS

The data are first presented separately for the two studies. For the adolescents, as can be seen from Table 8.1, the first four variables deal with knowledge about some aspects of the VCR (size of cassettes, type of video system, functions of buttons and off-air recording) while the second set of variables pertain to use of the VCR.

Overall, adolescent respondents were highly knowledgeable about the size of the cassette (93%) and somewhat less knowledge about how to record off-air (88%). The adolescents sampled showed moderate knowledge of the functions of the various buttons (78%) and little knowledge of the kind of video system owned by the family (31%).

As for the gender variable, there was a consistent difference indicating more knowledge among boys as compared with girls, with statistically significant difference for the type of system ($X^2 = 10.66$, $df = 1$, $p<.001$) and the functions of the buttons ($X^2 = 7.72$, $df = 1$, $p<.01$). As for the father's ethnic origin, there were mixed differences among the three subgroups, but the differences for the knowledge of the "size" of the cassettes was significant ($X^2 = 8.17$, $df = 2$, $p<.05$). Finally, as far as the longevity of owning the VCR is concerned, there was only a slight tendency for more knowledge about the system, the buttons, and off-air recording as the longevity increased.

Regarding VCR features, only moderate levels of use were reported: 51% for using the pause function, 61% for the picture search, and 77% claimed to re-view parts of programs. As for the breakdown by gender, there was essentially no difference between boys and girls except for slightly less use by girls of the search function. Regarding the ethnic origin, slight and inconsistent differences were found. Finally, regarding the amount of time owning the VCR, there was a tendency for less use of the various features as time went by, especially among those respondents who owned the VCR the most time.

As for the adolescents' perceptions of their parents' VCR knowledge and use, the respondents had dramatically different perceptions of their mother's and father's video know-how. Differences in the overall perception of fathers and mothers were calculated by means of related *t-tests* that were all significant at the $p<.05$ level or less. Respondents' fathers were perceived to know more about the functions of the VCR; some 61% of the fathers versus 14% of the mothers were perceived to know all the functions and 9% of the fathers and 34% of the mothers were perceived not to know any of the functions. Moreover, 92% of the fathers were perceived as being able to operate the machine by themselves, while only 73% of the mothers were perceived to be able to do so. Also, there was a great difference between the perceived ability of the fathers and mothers to record programs off-air (61% vs. 20%, respectively). Finally, 30% of the mothers were perceived to be "scared" of operating the VCR by themselves, compared to only 5% of the fathers.

The perceptions of the children regarding their parents were also examined separately for the boys and girls. Interestingly there were no significant differences between the perceptions of boys and girls. And

Table 8.2 Adolescents' Perception of Parents' Knowledge and Use of VCRs (in percentages) [n=83]

	Perception of Father by			Perception of Mother by		
	Total	Boys	Girls	Total	Boys	Girls
Parent knows function of *all* buttons	60.8	54.8	67.6	13.9	14.3	13.5
Parent doesn't know function of *any* button	8.9	7.1	10.8	34.2	35.7	32.4
Parent can operate VCR all by him/herself	92.2	92.5	91.9	72.6	70.3	75.0
Parent can program to record off-air	61.1	52.5	71.9	19.7	20.0	19.4
Parent is scared to use VCR by him/herself	5.3	7.7	2.8	30.0	40.5	18.2

yet, in three cases the differences, while not significant (probably due to the small number of cases), are notable: girls perceived their fathers more than boys did as knowing the functions of all the buttons (68% vs. 55%) and of being able to record off-air (72% vs. 53%). By contrast, boys were more likely than girls to perceive their mothers to be more frightened of using the VCR (41% vs. 18%).

When comparing the responses of the adolescents concerning themselves and their perception of their parents on the two relevant variables (knowing the functions of the buttons and being able to record off-air) it is clear that the children had greater expertise than their parents. Thus 78% of the children and 61% of the parents claimed to know what all the buttons are used for, and 88% of the children and 61% of the parents said they could record programs off-air.

The adults' sample, even to a greater extent than the parents-of-the-adolescents sample, showed that there was relatively little knowledge about the VCR and little use of its features. Only 70% believed they knew the format of the machine (although 92% indicated awareness of the more simplistic "size" of cassette characteristic). Also, only 41% said they knew the kind of system that they owned even though this figure was higher than for the children (31%). However, what is more striking is that only 45% of the adults claimed to know what all the buttons are used for (compared with 61% in the parents segment of the adolescents' sample). And finally, only 21% of the adults professed to use all the buttons on the machine.

Table 8.3 Adults' Responses by Gender, Age, Education, Ethnic Background, and Time Owning VCR (in percentages) [n=155]

	Total	Gender		Age		Education		
		Male	Female	−35	+35	−12yrs	12yrs	12+yrs
Know VCR format [VHS or Betamax]	69.7	83.1	50.8	81.9	56.3	59.4	67.3	77.9
Know size of cassettes	92.1	94.4	88.9	92.8	91.5	96.9	96.4	86.6
Know type of system [NTSC/PAL/SECAM]	40.8	48.9	29.0	42.2	39.4	34.4	28.6	54.5
Know function of *all* buttons[a]	44.8	57.1	27.0	54.8	31.9	40.6	43.9	46.3
Use *all* buttons[a]	21.1	26.7	12.9	21.7	19.7	25.0	19.6	19.7

	Ethnic Origin			Time owning VCR			
	Israel	West	East	−1yr	1-2yrs	3-4yrs	4+yrs
Know VCR format [VHS or Betamax]	74.5	75.8	45.8	69.2	69.6	71.0	74.4
Know size of cassettes	93.6	84.4	100.0	80.8	96.4	93.4	97.4
Know type of system [NTSC/PAL/SECAM]	42.1	50.0	21.7	42.3	41.1	29.0	48.7
Know function of *all* buttons	48.5	41.9	33.3	30.8	43.9	43.8	52.5
Use *all* buttons	22.1	22.6	16.7	12.0	22.8	28.1	17.5

[a]The data reported in the table pertain to the *all* category only while the values of X^2 reported in the text pertains to the complete spread of the variable.

As for differences based on demographic variables, it is clear that the men were significantly more knowledgeable of the VCR features compared with the women. Also, the men used them with greater frequency. The significant differences were obtained for the knowledge of the format variable ($X^2 = 16.79$, $df = 1$, $p < .01$), for the knowledge of the system variable ($X^2 = 5.99$, $df = 1$, $p < .05$), and for the knowledge of the buttons ($X^2 = 19.90$, $df = 4$, $p < .001$).

There was also a tendency for the younger adults (up to 35 years old) to claim to have better command of the machine compared with the older respondents (above 35 years of age). The findings were significant for knowledge of the format ($X^2 = 10.77$, $df = 1$, $p < .001$), for knowledge of the function of the buttons ($X^2 = 15.68$, $df = 4$, $p < .01$) and for use of the buttons ($X^2 = 15.87$, $df = 4$, $p < .01$).

As for educational level, the better educated were significantly more aware of the type of system ($X^2 = 9.17$, $df = 2$, $p < .01$) but not necessarily of the features, nor did they claim to use them more often.

Israeli born respondents and those from western countries were more aware of the format of the VCR than people from nonwestern origins ($X^2 = 8.12$, $df = 2$, $p < .05$) and used its features more often ($X^2 = 17.49$, $df = 8$, $p < .05$).

Finally, there was no clear relationship between longevity of VCR ownership and the knowledge variables except for the knowledge of the size of the cassettes ($X^2 = 8.62$, $df = 3$, $p < .05$). And yet there seems to be a tendency, albeit not significant, of knowing more of the functions as people own the machine longer, while at the same time the peak in terms of using the VCR features is at the midpoint in terms of longevity of ownership.

Conclusions

While the samples in the two studies reported above were relatively small (in part due to the need for much larger samples in order to locate sufficient VCR owners at a level of penetration of only 40%), and despite the fact that there was no perfect equivalence in the manner in which the variables were defined in both situations, the findings of both studies do seem to complement one another and to indicate that there is clearly less than perfect knowledge about the VCR and far from full utilization of its myriad of features. In this sense, then, it seems that these findings further reinforce the thesis that the VCR age has not brought with it a profound change in the use of mass media technology. For social scientists to accept the notion that a revolution has occurred, it would have been imperative to show that people know more about and actually utilize more of the functions and possibilities that the VCR has to offer.

Moreover, the consistent findings regarding the difference between men and women and between boys and girls seem to be in line with conclusions by Harvey and Rothe (1986) who found that the male head of the household maintained the most important role in all phases of VCR purchasing. Also, as Morley (1986) found in his study of 18 British families, "None of the women operated the video recorder themselves to any great extent, relying on husband or children to work it for them" (p. 158). Finally, Gray (1986) points out in the United

Kingdom that while some home appliances are perceived as "pink" and others as "blue"," in the case of VCRs one must distinguish between different operational elements: accordingly, some are "lilac" (the play, record, and rewind modes) but other functions which require presetting are considered "blue."

While it is often somewhat problematic to study media related issues and generalize from research across countries which differ in terms of their overall media ecology, it seems that these two studies, conducted in Israel, should not suffer from this dilemma. This would be the case because in the present research it was not the differential repertoire of available channels and contents that was under investigation, but rather how people relate to the technology of the VCR, which should not be any different from one location to another. It seems fair to imply, therefore, that people may indeed have big eyes (and perhaps also big appetites) for new technologies, but at least some of them also seem to have clumsy fingers.

REFERENCES

Boyd, D. & Straubhaar, J. (1985). Developmental impact of the home video cassette recorder on third world countries. *Journal of Broadcasting and Electronic Media, 29*, 5-21.

Cohen, A. A. (1987). Decision making in VCR rental libraries: Information use and behavior patterns. *American Behavioral Scientist, 30*, 495-508.

Cohen, A. A., Levy, M. & Golden, K. (1988) Children's uses and gratifications of home VCRs: Evolution or revolution? *Communication Research*, 15, 772-780.

Cubbitt, S. (1986). *Time shift: the specificity of video viewing*. Paper presented at the Television Studies Conference, London.

Gray, A. (1986). *Video recorders in the home: Women's work and boy's toys*. Paper presented to the 1986 International Television Studies Conference, London.

Gubern, R. (1985). La antropotronica: Neuvos modelos tecnoculturales de la sociedad mass-mediatica. In R. Rispa (Ed.), *Nuevas tecnologias en la vida cultural Española*. Madrid: FUNDESCO.

Harvey, M. G. & Rothe, J. T. (1986). Video cassette recorders: Their impact on viewers and advertisers. *Journal of Advertising Research, 25*, 15-19.

Jennings, M. K. & Niemi, R. G. (1974). *The political character of adolescence: The influence of families and schools*. Princeton, NJ: Princeton University Press.

Levy, M. R. (1987). Some problems of VCR research. *American Behavioral Scientist, 30*, 461-470.

Levy, M. & Gunter, B. (1988) *Home video and the changing nature of the audience*. London and Paris: John Libby.

Lin, C. (1987). A quantitative analysis of worldwide VCR penetration. *Communications: The European Journal, 13*, 131-148.

Lindlof, T., Shatzer, M. & Wilkinson, D. (1987). *Modes of accommodation of VCR-augmented television in the American family.* Paper presented at the meeting of the International Communication Association, Montreal.

Morley, D. (1986). *Family television: Cultural power and domestic leisure.* London: Comedia Publishing Group.

Schafer-Gross, L. (1986). *The new television technologies.* Dubuque: Wm. C. Brown.

9

Measuring VCR "Ad-Voidance"

BARRY S. SAPOLSKY
and
EDWARD FORREST

The 1960s and the 1970s witnessed a number of profound social movements—civil rights, antiwar, counterculture, and women's liberation. As for the 1980s, historians might consider another—the television viewer's liberation movement. For the first time in the history of the medium, a majority of viewers have taken control of what they watch, when they watch, and how they watch television. Indeed viewers throughout the world are uniting behind the liberating technology of the video cassette recorder. By tuning in, turning on, and dropping in a videocassette, viewers are no longer constrained by network schedules, limited channels, restricted choice, or commercial interruptions. Accordingly, as with any social movement, the established institutions are challenged and consternation and change are inevitable. In the U.S. the most threatening aspect of the VCR revolution grows out of its capacity to undermine the economic structure and logic of the commercial television industry.

Until the late 1970s, the multi-billion dollar assumption in television had been that if viewers watched the program, they also watched the commercials. As it is, viewers have always been able to register their disinterest in television commercials by inattention or leaving the room. Despite the fact that past research has indicated substantial levels of psychological and physical avoidance, this phenomenon has largely been ignored. However, with the advent of the home version of the VCR, the question of commercial avoidance takes on a new dimension. With the mere press of a button, viewers can zip past ads in programs they are watching that were previously recorded, viewers can delete ads entirely while recording, or ads can even be automatically deleted by a device called the Zapper, which electronically detects commercials during recording and edits them out.

The VCR was introduced to the consumer market in 1975, and it took four years to be in place in 1% of U.S. households (Lachenbach, 1983). It would be almost another six years before the VCR would reach the critical mass (a market penetration level of at least 33%)—the point at which media researchers regarded the VCR as a significant factor in audience measurement. A.C. Nielsen presently estimates that by the end of 1988, VCR penetration will be 60% of all U.S. households (Broadcast Marketing and Technology News, 1988).

Initial studies by Levy (1980, 1981) and A.C. Nielsen (1984, 1985) documented that the major function of VCRs was for timeshifting commercial programs. And, although these very same studies documented measurable and varying levels of VCR-facilitated commercial avoidance, advertising agencies and television network executives regarded the phenomenon of VCR *advoidance* as a minor audience aberration. However, television executives could hardly have found solace in Nielsen data that showed that over one-third of all VCR owners skipped virtually all the commercials during playback (Nielsen, 1985). Moreover, as the level of VCR adoption has escalated through the 1980s so too has the measured levels of advoidance. Indeed, the popularity of zipping appears to be growing dramatically. The 15% figure in 1979 (Levy, 1980), when the penetration level of VCRs was about 1%, appears to have grown to 50% in 1985 (Metzger, 1986) when the penetration level grew to 35%. Other recent studies have also found rates of commercial avoidance via zipping ranging from 50% to 70% of VCR owners (Forrest, Sapolsky, & Smith, 1986; Papazia, 1986; Reiss, 1986).

Subsequently, there has been no shortage of statistics, interpretation, and speculation regarding the impact of the VCR on commercial television viewership. One problem researchers of the phenomenon have noted is that, although the literature includes many estimates about past, present, and future rates of television commercial avoidance, they are widely divergent and consequently no one really knows yet the extent to which the VCR has impacted the audience for broadcast programs and for the advertising messages planted therein. Confusion over the nature and magnitude, as well as importance and impact, of commercial avoidance via the VCR can be attributed to at least three factors:

(1) Conflicting methods of measurement and operationalization of fast-forwarding and deletion of television commercials;
(2) Conflict of interest on behalf of the research organizations and practitioners who analyze and report the results of VCR studies that measure advoidance; and
(3) Conflicting and/or untested assumptions concerning the relationship between television viewers' attitudes toward advertising in general (and their evaluation of television advertisements in particular) and their pattern and likelihood of avoiding commercials when using their VCRs.

Conflicting Evidence and Measurement Techniques

Why is it, as one advertising agency media planner observed, that when it comes to measuring VCR commercial avoidance "Nielsen shows one number, agencies another, then the three conventional networks produce an independent report which shows a third set of figures?" (Triolo, 1986, p. 69). Potter, Wotring, Forrest and Sapolsky (1987) suggest that the logical starting place in sorting out the confusion is to review how television advertising avoidance has been defined and operationalized. First, there is confusion over the use of terms in the literature. Papazian (1986) makes a clear distinction between zipping and zapping, where zapping is commercial deletion during recording and zipping is fast forwarding during playback. Metzger (1986) makes a distinction between the two activities, but calls them both zapping. Reiss (1986) uses the term zapping but it is unclear if he includes deletions while recording along with fast forwarding. Forrest et al. (1986) make a clear distinction between deletion during recording and fast forwarding during playback (the latter they call zipping). However, they use the term zapping to refer to an non-VCR activity of avoiding ads during viewing by using a remote control device to change channels. Once the reader of this literature gets beyond the inconsistent use of terms, it is clear that these studies are primarily concerned with two activities of commercial avoidance: deleting of ads during recording and fast forwarding past ads during playback. In this study, these two activities are referred to as zapping and zipping, respectively.

A more serious shortcoming in the literature on VCR use is the way in which the concepts of zapping and zipping are measured. Some studies employ diaries wherein viewers provide self-reports of their

VCR use patterns over time. Other studies use one-shot surveys in which an individual in a VCR household is asked to indicate on a Likert-type scale his perceived general level of zipping. Typically, the respondent must choose from among the following alternatives: always, usually, sometimes, rarely, and never. Reiss (1986) reports on a study done for *Advertising Age* magazine using this type of response format which found that 47% of the respondents say they always zip and 25% say they never zip. The study found that commercials are zipped an average of 62% of the time. Forrest et al. (1986) report that 73% of VCR users regularly (summation of almost always, very often, and usually) zip past commercials during playback, and 48% regularly zap ads during recording. Papazian (1986) reports that 30% to 50% of VCR owners usually zap, while 50% to 60% usually zip. Metzger (1986) reports on a CONTAM VCR research study that used a slightly different method: asking people whether or not they zapped the previous day. One in twenty respondents was involved with zapping with the VCR.

There are two problems with self-report measurement of the concept of television advertising avoidance. First, comparisons across studies are suspect. In one study it may be reported that 50% of the respondents said they usually zip where "usually" is a summation of responses of always, almost always and very often. In another study, only 30% of respondents may say they usually zip but here "usually" is one response option on a five-point scale. Additionally, there is the interpretational problem across respondents because "usually" may imply a different frequency of occurrence to different people. This form of measurement makes it very difficult to assess the level of commercial avoidance and to track it over time.

The second, and perhaps more troublesome, problem with this type of measurement is that it does not tap the *absolute* amount of avoidance that is taking place—a measure that would be far more enlightening to advertisers. For example, the person who plays back 30 hours of programming a month and reports that she sometimes zips commercials will, ultimately, avoid more advertising than the person who says he always zips the ads during his two hours a month of VCR playback. Simply, it is not only a matter of how often respondents indicate they zap or zip. The level of videotaping activity must also be considered. Without taking into account how much time a person engages in playing back time-shifted programming, an accurate measure of the magnitude of advertising avoidance cannot be obtained.

Conflicts of Interest

A second source of confusion over the nature and magnitude of VCR commercial avoidance can be ascertained by even a cursory review of the literature, especially with respect to the research reported within the industry trade journals. Indeed, there often appears to be more than a modicum of interpretative spin put on the results of any VCR study or statistic. This is certainly understandable, given the fact that the commercial TV industry practitioners who supply the figures and commentary necessarily have a vested interest in the characterizations and conclusions which can be drawn from the advoidance data. This interpretative spin has included:

> Outright rejection of the statistics. It has been observed that "agency executives are quick to note that people don't always tell the truth in interviews ... there is a belief that viewers are not supposed to admit that they like to watch commercials." (Spillman, 1983, p. 82)

> Optimistic calculation: The VCR will only add to the existing commercial TV audience. As one advertising media planner suggested: "Growth of recording and playback could bring a trade off ... the additional viewers may turn out to be large enough to make up for the number of viewers lost to advertisers through skipping by their commercials ... In some cases it could even more than make up for the loss." (Swisshelm, 1984, p. 72)

> Negligible effect: In the final analysis VCRs will subtract very little of the commercial TV audience. As David Poltrack, Vice President of Marketing Development for CBS states, "We also would remind advertisers worried about this phenomenon that, even using Nielsen's suspect figures, zapping ["zipping" in the context of this chapter] currently affects approximately 0.4% of the average prime time program's audience and will only affect 1.1% of that audience when, and if, VCR penetration reaches 50%." (Christopher, 1985)

This last set of figures exemplifies the statistical sleight of hand possible through careful selection and manipulation of VCR advoidance data. An examination of the very same Nielsen data to which the network executive refers also indicates that program recording choice and commercial avoidance are not average phenomena. Nielsen data indicate that a limited number of programs account for the overwhelming amount of VCR playback and subsequent advoidance. Specifically, the A.C. Nielsen VCR Diary Study (1984) for the third quarter of 1984 reported that 41% of all VCR recordings involved only 11 different programs. Ten of these were daytime serials. Moreover, Nielsen also reported that 71% of the commercials contained in the recorded daytime serials were skipped during playback. An examination of prime-time programming data reveals that the most popular programs again account for the preponderance of VCR playback and, thus, commercial avoidance. When averaged out over all programs within or across all dayparts, which industry executives would prefer to do, the incidence of zipping will be necessarily diminished. However, advertisers on *The Bill Cosby Show*, who at the time were paying $380,000 for a thirty-second spot, might want a more accurate accounting of the delivered audience, in light of the fact that nearly 15% of all television households with VCRs were recording the program and probably zipping past some if not all of the commercials.

Untested Assumptions

Emerging findings on advertising avoidance have begun to directly challenge the critical assumption that viewers who watch the program also watch the commercials. Advertisers fear that as VCR penetration continues to increase, there will be a "loss of network ratings, lost revenue for advertisers, and further problems in accurate measuring" (Reitman, 1985, p. 86).

Advertisers were upset with the procedures used by the A.C. Nielsen Company that credited the taping of programs as viewing (*Broadcasting*, 1984). However, according to Nielsen's own reports, just 75% of recorded programs are ever played back and 60% of these involve the zipping of commercials during playback (Sternberg, 1988a). This approach of counting all taping activity as viewing while ignoring advoidance during playbacks was not initially a problem when VCR penetration was low and nearly all of the recording was done simultaneous to viewing of the same show. However, with increasing VCR penetration, advertisers are less accepting of the old methods. Kaplan (1985) points out that General Foods may be losing $1 million a year because of advertising nonexposure. With television ad rates increasing 50% faster than the overall rate of inflation over the last decade, advertisers are concerned about any development that would appear to reduce the effectiveness of their buys. Some ad agencies are even billing the networks for the lost audiences (Mandese, 1986). In response to the increasing nonplaybacks, A.C. Nielsen in early 1985 altered its measurement reports to show both conventional viewing and home recording of programs. However, Nielsen still cannot provide data which accurately reduce ratings for commercial exposure due to zapping and zipping.

Recently, network executives have begun to shift the burden of lost viewership through advoidance to the advertising practitioners by maintaining that it is their job to hold the viewers that the networks initially attract with programming. Advertising executives have responded with promises of "new and improved" creative strategies. As one advertising practitioner put it: "one thing we can do . . . . is to make commercials better . . . . enough better so people won't skip past them . . . . we want to put some entertainment or something into our messages so they can reach out and grab people and make them want to see them" (Swisshelm, 1984, p. 72).

This latter notion of making commercials irresistible appears to be the stratagem most professed by advertising agencies. Recent trade publications describe how the new television ads have been using more and bigger celebrities, humor, special effects, and music videos to compete for an audience containing a growing number of zippers and zappers. As the president of Ogilvy and Mather proclaimed: "The ultimate defense against being zapped is to involve the viewer. Today there must be some reason—a reward—for people to want to watch your advertising. That reason can range from information to entertainment" (Roman, 1984).

Concomitant with the copywriters' and creative directors' efforts to inject new television commercials with heightened levels of entertainment and information value is the American Association of Advertising Agencies' (AAAA) efforts to improve consumers' attitudes toward advertising in general. The efforts of the AAAA are motivated, at least in part, by the fact that "some 70 percent of Americans say they distrust advertising, finding it both offensive and an insult to their intelligence" (*Wall Street Journal*, 1986).

Other advoidance abatement strategies include the suggestions of Ronald Kaatz, formerly Senior Vice President of J. Walter Thompson: make "more exciting, more involving, more entertaining" commercials, and eliminate "commercial pods which are not surrounded by programming and are therefore more susceptible to zapping" (*Marketing News*, 1986, p. 12).

The critical assumption underlying industry efforts to improve advertising's content and image is that, if consumers regard advertising as more trustworthy, informative, and entertaining, and are more appreciative of its presence and intent, then they will be less likely to avoid advertising.

What is clear from a review of research on zipping activity is that measurement problems and statistical contrivances cloud our understanding of the nature of this form of advertising avoidance. An exploratory study of zipping behavior under highly controlled conditions was done to address some of the issues contributing to the confusion surrounding audience advoidance behaviors.

First, the critical measure of zipping activity is obtained through observation of actual VCR use. Rather than depend on self-reports of zipping, participants in a laboratory experiment are observed fast forwarding of their own volition. The exploration of zipping behavior also looks at the pattern of fast-forwarding activity across advertising pods or clusters embedded in popular TV entertainment fare. In addition, through comparison with self-report data gathered from the same participants, differences between reported and observed zipping can be assessed.

Finally, this study examines the relationship between advertising attributes and avoidance. Specifically, do unfavorable opinions of TV advertising in general correlate with the tendency to fast forward during commercial breaks? Furthermore, specific qualities were assessed for the advertisements used in this study. The relation of these evaluations to zipping activity is examined to test the assumptions of advertising practitioners.

METHOD

An exploratory investigation of VCR playback behavior, specifically fast-forwarding or zipping during commercials, was conducted as part of a larger study of television advertising avoidance. In general, the procedure involved exposing subjects to an episode of *The Bill Cosby Show* into which 16 pretested commercials were inserted. In one condition, subjects were able, at their own discretion, to fast-forward the videotape via remote control. Zipping behavior was observed by a video camera placed out of the subject's view. The amount of commercial material the subject avoided through fast-forwarding is the primary dependent variable. After viewing the videotape, subjects completed a questionnaire on which they reported, among other things, their advertising avoidance behaviors and attitudes toward television advertising.

The pretest was done in two stages. First, 166 different commercials being run nationally were observed in 18 hours of primetime network programming recorded from the three major networks (one hour of programming was obtained from each network during six consecutive evenings). These commercials were then rated in terms of familiarity by a sample of college undergraduates. The pretest subjects were given a list of brand names (from the 166 commercials) and asked to indicate if they recalled seeing a television commercial for that product brand in the previous two months. Their responses were made on a 10-point scale ranging from 1, "Do not recall," to 10, "Definitely recall." Average recall scores were rank ordered. The top 20 and the bottom 20 ads were selected yielding two groups: high familiarity and low familiarity.

The second stage of the pretest involved showing the 40 commercials to a separate sample of college undergraduates who rated the degree to which they found each commercial entertaining, informative, and believable. Subjects responded on 10-point scales ranging from 1, "Not at all," to 10, "Extremely."

An index of the three characteristics was created by summing the average rating on each characteristic. The index scores for all 20 high-familiarity and 20 low-familiarity commercials were then rank ordered. Within each level of familiarity, the four commercials with the highest indices and the four with the lowest indices were selected. The resulting 16 commercials were edited into the program material in four pods (clusters). Each pod contained one of each type of commercial (high-familiarity/high index, high familiarity/low index, low familiarity/high index, and low familiarity/low index).

The pods replaced the commercials that had accompanied the original airing of *The Bill Cosby Show* episode. The pods appeared right after the opening credits, approximately five minutes into the program, about 15 minutes into the program, and right before the closing credits.

The main experiment was conducted using two independent samples: (1) 29 male and 57 female undergraduates (hereafter referred to as *students*) enrolled in a communication course; and (2) 19 male and 16 female adult VCR owners (hereafter referred to as *adults*). Subjects were randomly assigned to one of three conditions: VCR remote control present, remote control present plus encouragement given to fast-forward, and control (no remote control present). Of interest in this study is the first condition: subjects free to choose to fast-forward through any portion of a previously recorded television program.

Under the guise of studying interpersonal behavior in television comedy, the subject was placed in a viewing room containing a television set, videocassette recorder, and, in the experimental conditions, a wireless remote control unit placed on a coffee table immediately in front of the subject. The subject was instructed to start the VCR by pressing the play button on the remote control device. The experimental assistant was situated in an adjoining room where he/she could observe the subject's fast-forwarding behavior through a television monitor connected to a video camera mounted on the wall above and behind the subject. The assistant registered the frequency and duration of the subject's zipping behavior using an event recorder. At the conclusion of the exposure phase, the subject was asked to complete a questionnaire regarding, among other things, VCR use and views on television advertising.

FINDINGS

Zipping Behavior

The level of fast-forwarding activity differs markedly between the student and adult samples. As can be seen at the bottom of Table 9.1, student subjects zipped an average 97.1 seconds or about one-fifth of the advertising content. In contrast, those in the adult sample zipped an average of 311.9 seconds, nearly two-thirds of the 480 seconds of commercial time contained in the experimental program. This difference may be explained in part on the basis of experience. A greater share of the adults in the experiment reported having used a VCR as well as a hand-held remote control for a VCR, television set and cable converter. Interestingly, a nearly equal number of adults (43%) and students (37%) indicated that the "ability to watch TV shows without the commercials" was important in their decision to purchase a VCR.

A greater number of subjects in the student sample chose not to fast-forward past the advertising content. Thirteen out of 28 in this sample did not zip any commercials. In contrast, only three out of 14 in the adult sample did not exhibit any zipping behavior. When the nonzippers are removed, the student sample averaged 181.2 seconds of fast-forwarding during ads. For the adult sample, this figure increased

Table 9.1 Zipping Behavior During Television Advertisements

Advertisement	Attributes[a]	Student Sample Average Seconds Zipped	Student Sample Percent of Adv. Zipped	Adult Sample Average Seconds Zipped	Adult Sample Percent of Adv. Zipped
Wendy's hamburgers	HH	2.89	9.6	14.14	47.1
Coor's light beer	HL	5.39	17.8	16.71	55.7
American Airlines supersavers	LH	5.11	17.0	17.21	57.4
Romegel antacid	LL	6.46	21.5	20.00	67.8
Total Pod 1		19.86	16.6	68.07	56.7
Wisk detergent	LL	5.79	19.3	16.43	54.8
Tucks medicated pads	LH	7.43	24.8	21.43	71.4
Bounce fabric softener	HL	5.61	18.7	20.36	67.9
Diet Coke	HH	2.71	8.0	17.50	58.3
Total Pod 2		21.54	18.0	75.71	63.1
Nyquil cold medicine	HL	5.32	17.7	19.64	65.5
Lysol laundry sanitizer	LH	9.53	31.7	20.71	69.0
American Express card	HH	8.25	27.5	20.00	67.8
Oldsmobile Cutlass	LL	9.03	30.1	21.93	73.1
Total Pod 3		32.14	26.8	82.29	68.6
Kraft cheese	LH	5.43	18.1	21.50	71.7
Suave lotion	LL	5.89	19.6	22.71	75.7
Pizza Hut Priazzo	HL	5.50	18.3	21.36	71.2
Alka Seltzer tablets	HH	6.71	22.4	20.21	67.4
Total Pod 4		23.54	19.6	85.79	71.5
Total All Advertisements		97.07	20.2	311.86	65.0

a. Attributes: First letter signifies level of familiarity (L = low; H = high). Second letter signifies appreciation index composed of ratings of "entertaining," "informative" and "believable" (L = low; H = high). Thus, a commercial with the attribute "HL" would have been rated in the pretests "high" on familiarity and "low" on the composite appreciation index.

to 396.9 seconds or 83% of the total advertising content. Thus, it would appear that, when adult VCR users fast-forward past television commercials, they skip a very substantial portion: as much as four out of every five seconds.

To determine if changes in fast-forwarding behavior occurred over the duration of the playback experience, a comparison of zipping behavior across the four commercial pods was done. As the program progressed, adult subjects engaged in increasing levels of zipping. Fast-forwarding jumped 26% or 17.7 seconds from the first to the fourth ad pod. This change was not significant, however (repeated measures analysis of variance yielded $F = 2.49, p = .07, df = 3,39$).

The pattern of zipping activity for the student sample remained level, except for the third ad pod where fast forwarding increased 8.8 seconds or 49% over the preceding pod. Change over time was found to be insignificant (repeated measures analysis of variance yielded $F = 1.92, p > .10, df = 3,83$).

Zipping and General Opinions of Advertising

On the postexperimental questionnaire five statements probed the subjects' perceptions of television advertising. Subjects were asked to indicate the degree to which they regard TV ads in general to be entertaining, informative, believable, annoying, and condescending. Responses were made on five-point scales ranging from Always to Never. The total number of seconds of advertising that was fast-forwarded was correlated with each of the advertising opinion scales. If, as in the view of advertising practitioners, a favorable predisposition toward television advertising will limit the viewer's inclination to fast-forward past commercials, then it would be expected that a negative relationship exists between fast-forwarding and the belief that ads are entertaining, informative, or believable. By the same token, a positive relationship between zipping and views that TV ads are annoying or condescending should emerge.

It is assumed that the viewer is able to recognize enough of a fast-forwarded commercial for information processing to take place, including recollection of affective response during previous exposure to the commercial. Nearly all VCRs allow a reasonably clear image to appear during fast-forwarding which should permit the viewer to determine the content of an advertisement (however, the audio component is unavailable during fast-forwarding).

None of the relationships between zipping behavior and opinions of television advertising obtained significance for either the adult or the student sample. For the adult sample, the direction of the correlations is consistent with expectations except for the attribute "entertaining" ($r = .334$). For the student sample, the direction of the relationship between zipping and the qualities of "believable" ($r = .223$) and "condescending" ($r = -.008$) is contrary to expectations.

In general, the absence of a significant relationship between observed zipping activity and opinions of TV advertising is consistent

Table 9.2 Average Seconds Zipped for Commercials Varying in Familiarity and Appreciation

	Hi Familiar Hi Apprec.	Hi Familiar Lo Apprec.	Lo Familiar Hi Apprec.	Lo Familiar Lo Apprec.
Students	5.14_a	5.46_{ab}	8.66_c	6.79_b
	$F = 5.05, p < .01, df = 3,81$			
Adults	17.96_a	19.52_a	28.29_b	20.27_a
	$F = 23.51, p < .01, df = 3,39$			

NOTE: Means having different subscripts differ significantly at $p < .05$ by t-test.

with earlier work limited to *reported* avoidance behavior (e.g., Forrest et al., 1986).

Zipping and Pretested Advertising Attributes

Every commercial inserted into the experimental videotape was rated in a pretest on the attributes of entertaining, informative, and believable. A composite appreciation index was developed using these three characteristics. Half of the 16 commercials imbedded in the experimental videotape were classified as "high" on the appreciation index and half as "low" on the index. An additional advertising attribute was measured during pretesting: an advertisement's familiarity. Again, half of the experiment's 16 commercials were classified as high familiarity and half as low familiarity. Thus, four categories of commercials were utilized: (1) high familiarity/high appreciation index, (2) high familiarity/low appreciation, (3) low familiarity/high appreciation, and (4) low familiarity/low appreciation. A separate repeated measures analysis of variance was conducted on fast-forwarding behavior for each sample. As can be seen in Table 9.2, significantly more zipping activity occurred in both samples in the condition of advertisements that were rated both low on familiarity and high on the appreciation index.

To further clarify the contribution of familiarity and the components of appreciation to zipping behavior, total seconds zipped were compared for ads rated high and ads rated low on each characteristic.

Table 9.3 Average Seconds Zipped for Commercials Differing on Pretested Levels of Entertainment, Informative, Believable, and Familiarity

Attribute	Student Sample		
	Average Seconds Zipped	t value	Significance
High entertainment	5.56	1.36	$p > .05$
Low entertainment	6.46		
High informative	5.87	0.58	$p > .05$
Low informative	6.26		
High believable	5.95	0.79	$p > .05$
Low believable	6.21		
High familiarity	5.30	2.41	$p < .05$
Low familiarity	7.73		

Attribute	Adult Sample		
	Average Seconds Zipped	t value	Significance
High entertainment	18.77	0.14	$p > .05$
Low entertainment	20.06		
High informative	19.35	0.46	$p > .05$
Low informative	19.63		
High believable	19.34	0.52	$p > .05$
Low believable	19.68		
High familiarity	18.74	5.06	$p < .01$
Low familiarity	24.28		

Table 9.3 presents the results of these comparisons. For both samples, the amount of zipping engaged in during commercials rated highly in terms of entertainment, informativeness, and believability does not differ from the level of zipping during lower-rated ads. This finding lends further support to the notion that television viewers will use the technology to avoid advertising, regardless of how much they enjoy and are informed by the advertising.

Total zipping behavior during high familiarity commercials was compared to that during low familiarity commercials. For the adult sample, high familiarity ads were zipped significantly less than low

Table 9.4 Relationship Between Observed and Reported Zipping Behavior

Reported Avoidance Behavior	Correlation[a] with Total Seconds of Observed Zipping	
	Student Sample	Adult Sample
Leaving room (skipping)	.321*	.097
Changing channels (grazing)	.080	.211
Turning off sound (muting)	.217	.297
Removing ads during recording (zapping)	−.221	−.062
Fast forwarding during playback (zipping)	−.063	.369

a. Pearson Product Moment Correlation. Correlation marked with an asterisk (*) is significant at $p < .05$. All other correlations do not reach significance.

familiarity ads. A similar difference was in evidence in the student sample (see Table 9.3).

Observed and Reported Avoidance Behavior

Subjects were asked how often they: (1) leave the room during commercials (skipping); (2) switch channels during ads (grazing); (3) turn off the sound during commercials (muting); (4) delete ads while recording programs on a VCR (zapping); and (5) fast forward past commercials during VCR playback (zipping). Their responses could range from Almost always to Never. Responses to these self-reports of advertising avoidance behavior were correlated with observed zipping behavior during the experiment. As shown in Table 4, observed zipping is not significantly related to reported avoidance behaviors for the adults. The correlations are, as expected, positive except for zapping behavior. Here, the less adult subjects reported they zap ads, the more they were found to zip them during the experiment. This would in fact be expected: if a VCR user does not routinely zap commercials during recording, then they are available to be zipped during later playback.

For the student sample the degree to which they report leaving the room during commercials is significantly related to zipping activity. No other correlations were significant. As with the adult sample, reports of zapping are negatively related to observed zipping. Of interest, for the students reported zipping behavior is negatively, albeit trivially, related to actual zipping.

Discussion

Observed zipping behavior among adults substantiates the concerns of the television industry: A large share of viewers avoid much of the advertising accompanying television programming. Two-thirds of the advertising content was fast-forwarded by the adult sample. More importantly, when nonzippers, the minority of viewers who choose not to avail themselves of the technology, are excluded, four out of every five seconds of commercial material are avoided. College students exhibit far lower levels of zipping activity.

As was observed, as the program progressed through commercial pods there was a tendency for adults to engage in increasing amounts of zipping. It would be expected that zipping will increase as the program progresses, due to the viewer's escalating involvement in the storyline and the desire to avoid the distraction and delay brought on by commercial breaks. Zipping activity continued to increase during the final ad pod *after* the conclusion of the episode. This could possibly be the result of a zipping "reflex" (a pattern of behavior built up over the course of TV viewing). Simply, fast-forwarding is easy, convenient and rewarding, and a tendency may develop to avoid all advertising breaks in this way.

The student sample increased zipping behavior only during the third of four advertising pods. According to the reasoning above, it would be expected that zipping would increase at this point (third pod). However, why zipping behavior was lower before and after the third advertising pod remains unclear.

Viewer attitudes toward television advertising in general show no relationship to observed zipping behavior. Individuals who like or abhor commercials are equally likely to zip past them while playing back prerecorded programs. This suggests that if individuals have the new video technologies at their fingertips, a sizeable share of them will utilize these technologies to avoid commercials regardless of how favorably or unfavorably they perceive television commercials.

When considering attributes of the specific commercials utilized in the present study, only degree of familiarity affected advertising avoidance. Television commercials that are more easily recognized as a result of prior exposure apparently have the capacity to involve the viewer, reducing avoidance significantly. Moreover, an advertisement's familiarity may be a more important factor in zipping behavior than its affect-eliciting qualities. Commercials rated highly in terms of being entertaining, informative or believable were zipped as frequently as those rated lower on these qualities.

Finally, *observed* zipping behavior was found to be unrelated to several *reported* advoidance behaviors. While nearly all of the correlations were found to be in the expected direction, the general lack of significant relationships suggests the limitation of self-reported avoidance behaviors. Alternatively, zipping activity in the artificial setting of the laboratory may be exaggerated.

Regardless of how the phenomenon of commercial avoidance is defined and measured, one thing is certain: The future will bring increasing penetration of new video technologies. With them will no doubt come ever greater levels of zapping and zipping. With the increase in commercial avoidance via the new technologies, advertisers will even more vigorously question the network-station-agency logic of basing spot costs on program ratings. The new video technologies threaten commercial telecasting by rendering its traditional physical advantages meaningless. In the past, through simple inertia viewers tended to remain in their chairs and take in the commercials with their programs. Even if they did leave the room, the audimeter kept running. As it was, everything worked in the broadcaster's favor. Audience measurement techniques plus the viewers' inertia combined to ensure that one could count *all* of the program viewers as watching *all* of the commercials *all* of the time.

Results of this and other studies of audience viewing patterns in the new video environment are directly challenging such assumptions. The fact that the structure and assumptions of the commercial television industry are in a state of flux is clearly exemplified in the ongoing debate between the advertisers, their agencies, the commercial networks and the industry's audience measurement services. For instance, one researcher suggests:

> The problem here is not only the obviously different interests of the networks and advertisers but also the definition of Nielsen's basic function in the marketplace and that of audience measurement services in general. The network position seems to be that since Nielsen is not a commercial measurement system, and won't be for some time, the program content should be included in the ratings regardless of whether all the commercials are "zipped" through. (Sternberg, 1988b, p. 123)

Advertising practitioners have a markedly different view of Nielsen's function in the marketplace:

> Neilsen came into being as a service for advertisers at a time when network programming was fully sponsored and advertisers virtually controlled programming decisions. As a service to advertisers, Nielsen is not in the

business of program measurement. *Nielsen's essential function is to measure the opportunity to see the commercial within the program.* (Sternberg, 1988b, p. 123)

As it stands, the emergence and continued proliferation of VCRs raises not only immediate implications concerning advertising rates, but also important methodological questions concerning audience measurement. Perhaps, in addition to the traditional demographic profiles, tomorrow's sponsors will require technographic profiles as well. Such technographic profiles would focus on both the ownership and the centrality that new information technologies play in a consumer's life-style and media-style. As this study has demonstrated, the mere presence of a VCR is enough to influence a consumer's television viewing or, more specifically, a consumer's TV commercial *non*viewing behavior, regardless of his predispositions toward advertising in general or the informative and entertainment value of any given commercial.

The present study is an initial attempt to examine commercial avoidance under controlled conditions. The measurement problems associated with self-reported zipping behavior are overcome by observation of actual fast-forwarding behavior under simulated home-viewing conditions. However, conclusions regarding observed zipping must be tempered by several methodological limitations. First, fast-forwarding may have been induced by demand characteristics—subjects were placed in an experimental situation with a remote control device (RCD) and VCR but with no other competing activities or distractions typically available in the home. Without the opportunity for conversation, snacking, reading, etc., viewers were only left with the choice of viewing and using the RCD. Future observational research on VCR advoidance should provide a more realistic viewing situation, which may of course provide even more avenues for advertising avoidance.

The choice of commercial content is a critical matter in research on zipping. Attributes such as the entertainment potential of advertisements should be assessed by individuals as similar as possible to those participating in experiments on zipping. It should be kept in mind that in the present study the ads were pretested using a *student* sample. Those in the adult sample might rate the advertising differently on the above-mentioned qualities. The experiment did not include evaluations of the 16 commercials by the adult sample. The use of commercials that have never aired is one means of preventing differences in prior exposure to ads—a factor that appears to affect zipping activity.

Lastly, larger samples of adults willing to participate in experiments on television viewing behavior are essential to definitive tests of advoidance. Media experiments have always been fraught with reliance on college student samples. The present study clearly shows that zipping is substantially different in adults and college students. Future research on advoidance will benefit by both a departure from self-report measures and utilization of more representative samples of the viewing public.

REFERENCES

Broadcasting (1984). Nielsen to break out VCR viewing. (November 26): 46.

Broadcast and Technology News (1988). National Association of Broadcasters Info-Pak. (July-August): 1-7.

Christopher, M. (1985, June 10) Net's audience may be widened by VCR use. *Advertising Age,* 60.

Forrest, E., Sapolsky, B. S., & Smith, E. (1986). *Advertising avoidance via RCDs and VCRs and attitudes toward TV advertising.* Paper presented to the Annual Convention of the Broadcast Education Association, Las Vegas.

Kaplan, B. (1985). TV commericial zapping—The issue is communication. *Marketing & Media Decisions, 20,* 100-102.

Lachenbach, D. (1983). Home video: Home is where the action is. *Channels of Communication, 3,* 42-43.

Levy, M. R. (1980). Program playback preferences in VCR households. *Journal of Broadcasting, 24,* 327-336.

Levy, M. R. (1981). Home video recorders and time shifting. *Journalism Quarterly, 58,* 401-405.

Mandese, J. (1986). Bates bullies nets over VCR erosion. *Adweek, 27,* 1, 4.

Marketing News (1986, May 9). Advertisers need quick fix, zipping, zapping. *20,* 12.

Metzger, G. (1986). Comtam's VCR research. *Journal of Advertising Research, 26,* 8-12.

Nielsen, A. C. (1984) Homevideo index: Video cassette recorder usage report. (July-September).

Nielsen, A. C. (1985). Homevideo index: Video cassette recorder usage report. (October-December).

Papazian, E. (1986). Zapping: Not just a media problem! *Marketing & Media Decisions, 21,* 103-104.

Potter, J., Wotring, E., Forrest, E., Sapolsky, B. (1987). *Zapping and zipping: Comparing alternative operationalizations.* Unpublished paper. Florida State University.

Reiss, C. (1986, October 27). Fast-forward ads deliver. *Advertising Age, 3,* 97.

Reitman, J. (1985). VCR's: The saga continues. *Marketing & Media Decisions, 20,* 83, 86-88, 156-158.

Roman, K. (1984, October 1). One buyer's opinion: Learning to live with the zap. *Television/Radio Age,* 69.

Spillman, S. (1983). Is TV spot zapping zooming? *Advertising Age, 54,* 1, 82.

Sternberg, S. (1988a). VCRs: Impact and implications. *Marketing & Media Decisions*, 22, 100-107.

Sternberg, S. (1988b). Measuring VCR playback. *Marketing & Media Decisions*, 23, 122-124.

Swisshelm, G. (1984, August 6). Ad zapping may be minor now, but major problems lie ahead. *Television/Radio Age*, 38-39, 72.

Triolo, P. (1986, March 31). VCRs no cause for panic, says Esty media director. *Television/Radio Age*, 69.

Wall Street Journal (1986, January 9). Advertisers launch an ad campaign to improve image. 1.

10

School Achievement, Self-Esteem, and Adolescents' Video Use

KEITH ROE

This chapter analyses the VCR use of adolescents in relation to their school achievement and self-esteem. In order fully to understand the uses made of VCRs by Swedish adolescents, it is first necessary to set that use within the context of the pattern of diffusion of VCRs in Sweden and the debate that accompanied it. The analysis is based on the results of a survey, funded by The Bank of Sweden Tercentenary Foundation, the purposes of which is to map the media use of Swedish adolescents. Data were collected from 1,334 fifteen year olds, divided between city, small town, and rural areas.

In Sweden, sales of VCRs began to increase in 1980 and then accelerated rapidly in 1981, in which year the percentage of households containing a VCR increased from 3.5% to 8% (Hulten, 1985). Subsequently, household penetration reached 16% in 1983, 23% in 1985, and 41% by early 1988 (Gahlin & Nordström, 1988). The main explanation for the comparatively early market take-off of VCRs in Sweden is probably to be found in the media structure of the country. Swedish television consists of only two channels of the public service type, and cable-TV, introduced in 1983, still reaches only about 13% of the population (Nordisk Medie Nyt, 1988: 3).

The advent of the VCR was accompanied by an intense public debate. The fact that the VCR facilitated the circumvention of TV and cinema censorship led to its utilization for viewing hired cassettes portraying explicit violence and pornography. In this way the VCR came to symbolize the negative aspects and consequences of new media technology and the threat being posed to traditional culture. In particular, the VCR was purported to be a threat to children and adolescents and a stimulus to aggresiveness and violent behaviour. A major campaign developed, which embraced parent-teacher associations, academics, moral campaigners, and politicians. In due course, virtually all of the criteria for a "moral panic" (Cohen, 1980; Roe, 1985) were fulfilled, and the VCR became something of a "folk devil" (ibid).

In response to these developments, the government instituted an official inquiry; a tax was imposed on the purchase of VCR equipment and measures were taken to enable the prosecution of dealers purveying cassettes containing explicit violence. The official inquiry also recommended (Ds U 1987: 8; S.O.U. 1988: 28) that the penalties for illegal distribution of violent content be stiffened and that the distribution of prerecorded home video cassettes be supervised by the National Film Institute. However, a proposal to subject all prerecorded home video cassettes to censorship control was rejected on the grounds that such a measure would be ineffective.

VCR use is widespread among adolescents. Only 15% never use a VCR, with females clearly outnumbering males in this group. On weekends as many as 43% of those with home access to a VCR utilize it for viewing. However, among the remainder use tends to be infrequent. Among regular users, males and those from lower socioeconomic status (SES) households are clearly overrepresented. The average time spent using VCRs by those with home access is 7 1/2 hours a week, compared to 2 1/2 hours for those without home access. An extreme group of 3% watches over 20 hours a week. Overall, detective/police stories, adventure films, and horror films are viewed most frequently, although there are large differences here according to gender. In one study of 12-16 year olds, 67% of the film titles named as having been seen on video were classified as only permissible for (cinema) viewing for those aged 15 or over; while 7% were banned from the cinema altogether. Horror films and films characterized by significant amounts of explicit violence constituted 39% of all the films named as having been seen (Roe & Johnsson-Smaragdi, 1987; Gahlin & Nordström, 1988; Wall, 1988).

Adolescence is generally regarded as a phase characterized by, amongst other things, strivings for greater autonomy. This manifests itself, not least, in shifts in media use and other leisure patterns. However, as Willis (1978) has stressed, existing media are likely to be "policed" by the dominant (adult) culture. Nevertheless, a new medium provides certain groups with opportunities for appropriation and use not anticipated or comprehended by the agents of the dominant culture, thereby creating a temporary space within which a group may act until the policing agencies of the dominant culture move to control the medium more effectively. The way in which some groups of adolescents in Sweden utilized the VCR in order to bypass cinema censorship and control may be seen as an excellent illustration of this process (see also Roe, 1983a; Svendsen & Vilsvik, 1987).

Early studies among adolescents (e.g., Roe, 1981) found that the VCR was being used largely to facilitate peer group interaction. In this respect it appears to have differed from the British and the U.S. pattern of VCR use, which is characterized by comparatively privatistic, individual behavior (Gunter & Levy, 1987; Levy, 1987). Furthermore, only a small percentage were found to be using VCRs with any frequency for time-shifting TV programs. Instead, the "home cinema" type of use predominated. More recent studies confirm that while the incidence of time-shifting has increased substantially, Scandinavian adolescents continue to use VCRs mostly for viewing prerecorded cassettes in the company of peers (Höjerback, 1986; Svendsen & Vilsvik, 1987).

Some group-based VCR use also manifests subcultural characteristics. One important and often overlooked factor in subcultures is the meaning that school has for adolescents. It is significant that most subcultures tend to occur during the final years of compulsory school when the meaninglessness of school for many adolescents leads to alternative investments in deviant or delinquent subcultures in which leisure and freedom from restraint are to the fore (Mann, 1981; Roe, 1987a). The crucial condition for the emergence of subcultures is the existence of a number of actors, in effective interaction with one another, with similar problems of adjustment, who attempt collectively to resolve these problems. In this process forms of collective identity are generated from which an individual identity can be achieved outside of that ascribed by class, education, and occupation. Insofar as a subculture represents a new status system, sanctioning behavior disapproved of by the larger society, the acquisition of status within the group is accompanied by a loss of status outside of it, which may, in turn, result in mutually hostile and contemptuous images. Empirically, subcultures tend to be located in areas of the social structure that provide a temporary escape from the authority and control of the dominant culture (Cohen, 1970; Brake, 1980).

In Sweden, school achievement has been found to be a good predictor of adolescents' music preferences (Roe, 1983b), music television use (Roe & Löfgren, 1988), and VCR use (Roe, 1987b). In the last-named study, based on data obtained from the same respondents who form the basis of the analysis in this chapter, significant negative correlations were obtained between school achievement and frequency of use of films featuring explicit violence, karate, horror, pornography, and police stories. In other words, those with lower school achievement viewed these categories more frequently than did those with higher school achievement. Lower school achievement was also found to be significantly related to greater home access to a VCR, to a higher level of use, to viewing in a larger group, and to a pattern of use in which prerecorded cassettes rather than time-shifted TV programs constituted the predominant viewing fare. These results suggest that, against the background of the highly charged moral debates which surrounded its arrival, some low achieving adolescents may have appropriated the VCR for socially disapproved of uses and as a means of expression for antischool subcultures (see also Roe & Salomonsson, 1983; Svendsen and Vilsvik, 1987).

Besides the struggle for autonomy, most observers are agreed that the search for esteem, approval, and identity is a basic human need (Maslow, 1968) and a primary concern of adolescence. While this process is "located" in the individual, it is also central to the individual's collective culture (Erikson, 1968). It is through their negotiations with family, peers, and demands for achievement that adolescents, individually and collectively, acquire the bases of their social world. It is also through these negotiations that a structure of gratifications needs is created that influences social action and the selection of media forms and contents. It follows that those experiencing difficulty in these respects may be led into a search for the self-esteem and identity not provided by legitimate social agencies and, in association, a search for alternative gratifications may be stimulated (Dembo, 1972, 1973).

Identifying the social circumstances that lead people to turn to the media for the satisfaction of certain needs has always been a central concern of uses and gratifications research and it has been postulated that media use is determined by the interaction between psychological dispositions, sociological factors, and environmental conditions (Katz et al. 1974). Following this, we can postulate that the use made by Swedish adolescents of "undesirable" prerecorded video cassettes is determined by the need to sustain self-esteem in the face of denigration resulting from failure or other negative experiences in school, in the context of the highly charged moral debate surrounding the arrival of the VCR in Sweden.

Before proceeding to an analysis of the relationships between school experiences, self-esteem and adolescents' VCR use, a brief review of existing research into the relationship between school experience and self-esteem will be presented.

In the development of self-esteem, many researchers give primary weight to parental treatment rather than to school experiences, claiming that the predispositions toward achievement and underachievement are present before children come to school (Purkey, 1970). However, while denying neither the importance of parental treatment, nor its temporal precedence, it can also be contended that the school is an arena of vital importance for the development of self-esteem. Consequently, if school experiences fail to create the necessary conditions for the development of self-esteem, then adolescents may turn, individually or collectively, to the media for alternative sources of identity.

The self system is an organization of experience for avoiding increasing degrees of anxiety that are connected with the educative process (Sullivan, 1953: 166; see also Rogers, 1951). Students' perceptions of self and the world are not only products of how others see them, they are also primary forces in academic achievement (Purkey, 1970). One of the most important single assumptions of self theories is that the maintenance and enhancement of the perceived self is the motive for all behavior. It follows that not only is the self the basic frame around which the rest of the perceptual field is organized, being both a product of experience and a producer of whatever new experience the individual is capable of, it is also essentially conservative and actively strives to maintain itself (Purkey, 1970). This means that where opportunities for high self-esteem are lacking, and where low self-esteem contributes to a negative self-identity, behavior will be directed toward maintaining and enhancing that negative identity. This helps to explain why some adolescents persist in deviant or deliquent behaviour—even in the face of the disapproval and punishment of the dominant culture. In short, a negative identity is better than no identity at all.

Loss of identity is often expressed in a scornful hostility toward the roles offered as proper and desirable in one's family or immediate community. In some cases, "a negative identity is dictated by the necessity of finding and defending a niche of one's own against the excessive ideals demanded by morbidly ambitious parents" (Erikson, 1968: 172). Vindictive choices of a negative identity then represent an attempt to regain some control over a situation in which the available positive identity elements cancel each other out.

In an Australian study, Power (1986) found that many adolescents saw teachers as a source of rejection or denigration, and were acutely aware of information or circumstances that might damage their self-esteem, as well as of sources of more favorable information about their competence as human beings. The rejected inevitably draw together and form an alternative subculture with its own definitions of the situation and rules of membership. Some students, in order to preserve a sense of identity in what is seen as an alien environment, become affiliated with an antischool group (see also Hargreaves, 1967; Elder, 1968; Roe, 1983b; Schostak, 1986).

Existing research indicates a clear relationship between self-esteem and school achievement, although the actual pattern is somewhat complicated by gender differences. While male underachievers tend to have lower self-esteem than female underachievers, successful females tend to have lower opinions of their academic ability, and themselves, than do comparable male groups. Furthermore, there appears to be dispute concerning whether, overall, males or females report a higher self-esteem. In general, while high achievers tend to be characterized by self-confidence, self-acceptance and higher self-esteem, low achievers are more likely to be depressed and anxious and tend to be characterized by a sense of incompetence, a preoccupation with self, and a feeling that others neither really understand nor respect them. In their relationships they are likely to elicit responses that confirm their apprehensions and feelings of rejection (Rosenberg, 1965; Wylie, 1968; Purkey, 1970).

The question of causal direction in the relationship between self-esteem and school achievement is a matter of dispute. A review of the available literature seems to indicate a continuous reciprocal relationship, not infrequently involving a self-fulfilling process whereby poor grades lead to a loss of self-esteem, a rationalization of one's performance and, ultimately, to hostility and dissatisfaction with the course, the teacher, and the school. This encompasses the neglect of study and the devotion of time to other activities (e.g., the media) — causing a further decline in achievement. Conversely, success raises self-regard, usually fairly permanently, and sets in motion a positive chain of development (Stinchcombe, 1964; Elder, 1968). Poor grades also tend to lower teacher expectations of students, these are then internalized by students into self-perceptions which influence future attainment and, once again, a self-fulfilling prophecy is set in motion (Rosental & Jacobson, 1968).

Since low self-esteem may be both a cause and an effect of low grades, it follows that the choice of causal direction in the relationship is largely a matter of perspective and is of less significance than analyses of the interaction between these variables and the consequences of that interaction for behavior. Such a view is supported by Bandura's (1978:345) contention that psychological functioning involves continuous reciprocal interaction between behavioral, cognitive, and environmental influences, and by Salomon's (1981) interactional approach which emphasizes spiral, rather than linear, cause and effect relationships.

The dynamic interaction between education and communication is best amplified by Bourdieu (1967, 1977, 1979, 1984). He argues that the various experiences of schooling provide recipients with common thought categories that provide dispositions for the reception of the messages of the culture industry. In this way education can be said to structure the media audience. It is always important to remember that people do not come to the media as blank slates, nor do they (or the media) exist in a social vacuum. They come with individual expectations and needs, as members of social groups, and with socially structured beliefs, knowledge, dispositions, and world views.

These pre-existing constraints not only influence interpretation and perception, they also influence whether or not people expose themselves at all to particular media contents. It follows that media choices are determined, at least in part, by what people "perceive to be appropriate, valid, informative, entertaining, or useful in the light of their prior experience and anticipations" (Salomon, 1981: 81).

Very little research has directly addressed the relationship between personality variables and VCR use, and those that have done so have reported conflicting results (e.g., while Greenberg and Heeter, 1987 found no consistent personality differences between VCR and non-VCR youth; Svendsen and Vilsvik, 1987 report significant relationships between self-concept and VCR preferences). Nevertheless, we hypothesize that both school achievement and adolescents' perceptions of self, separately or in interaction, will enable us to predict the uses which adolescents make of the VCR, and the gratifications they obtain from that use.

The remainder of this article will be devoted to an analysis of these relationships. Furthermore, in order to locate these relationships within the social context, the "background" variables of fathers' occupation, mothers' education, and gender will be included in the analysis. Finally, since much VCR use appears to manifest peer group and subcultural characteristics, adolescents' level of peer group activity and relative orientation to parents and peers will also be dealt with.

METHOD

The data presented here were collected by questionnaires administered in the classroom to 1,334 adolescents aged 15-16 years. They include one ninth-year class in virtually every school in the city of Malmö (population 240,000); every ninth-year student attending normal schools in the town of Kristianstad (population 55,000) and in three rural communities (population 40,000). In Sweden the ninth year is the final year of compulsory schooling, although most students continue with some form of education after its completion. Overall, a response rate of 98% was achieved.

By means of access to school records, the grades awarded to each respondent at the end of the seventh year, the eight year, and the autumn term of the ninth year were collected. The end of the ninth-year grades were awarded three to four months before the administration of the questionnaire. In Swedish schools, for each subject studied, each student is awarded a grade on a scale from one to five, where one represents the lowest possible grade, three the average, and five the highest. For the purposes of the analysis presented here, the overall average of the grades awarded to each student in all subjects was employed. Socioeconomic status data on parents' occupation and formal education were collected from the respondents and, wherever possible, public register data.

In addition to school achievement, adolescents' commitment to school has been found to be significantly related to their media use (Murdock & Phelps, 1973; Roe, 1983b). In this study, school commitment was measured by a scale adapted from Murdock and Phelps (ibid). Respondents were required to express the extent to which they agreed or disagreed with the following statements:

(a) On the whole I am pretty happy at school.
(b) At school, a lot of lessons are really a complete waste of time.
(c) School is always the same.
(d) At school, they nearly always treat you like a kid.

(e) After the holidays I am usually pretty happy to go back to school.
(f) A lot of teachers don't really try hard enough to make the lessons interesting.
(g) I can't wait to leave school and start work.
(h) At school, I am able to think for myself and use my imagination.
(i) Most teachers don't really care about us pupils.
(j) School is a good preparation for life.

Furthermore, by means of the following question, respondents were asked to grade their school: "The school gives you a grade every year—now you have a chance to give the school a grade. What grade do you want to give your school?" The answer alternatives were 1, 2, 3, 4 and 5, thereby corresponding to the scale applied by the school when grading the pupils.

Finally, respondents were asked to indicate how much homework they usually did.

Self-esteem was measured by a scale adapted from Rosenberg (1965). Respondents were asked to express the degree to which they agreed or disagreed with the following statements:

(a) On the whole I am satisfied with myself.
(b) Sometimes I feel completely worthless.
(c) I think that I have a lot of good qualities.
(d) I am able to do things as well as my friends.
(e) I wish I could like myself more than I do.
(f) It's quite easy for me to feel a failure.
(g) I find it difficult to speak in front of the whole class.
(h) I give up easily when things get difficult.
(i) It's easy for me to talk to other people.
(j) My teachers often make me feel that I am good for nothing.
(k) I am satisfied with my looks.
(l) I would change myself a lot if I could.
(m) I am actually satisfied with being a boy/girl.

An exploratory factor analysis indicated that items (g) and (i) did not adequately fit into the factor structure. It would appear that speaking in public is a situation provoking anxiety in ways not directly related to other aspects of self-esteem. Consequently, these items have been excluded in the subsequent analysis.

The self-esteem data were complemented by a number of items designed to measure respondents' feelings about life in general. These were:

(a) I think I often get the blame for things I haven't done.
(b) My parents often disapprove of my friends.
(c) I think there is a lot in life to be happy about.
(d) I often feel lonely and exposed.
(e) Instead of getting better, I think everything in life is getting worse.
(f) People don't really care about what happens to others.
(g) I often feel that life is meaningless.
(h) I often feel bored.
(i) Everything feels so uncertain and insecure nowadays.

Frequency and amount of viewing of VCRs was decided between weekdays (Monday to Thursday) and weekends (Friday to Sunday). On the basis of this data, three variables were constructed: amount of VCR use per week, per weekday, and per weekend.

Respondents were also asked to indicate the frequency with which they use a VCR to view the following: detective/police stories, violent films (of the kind certified or censored at the cinema), karate, horror films, adventure films, and pornography.

Finally, as indicators of the extent of their peer group involvement, respondents were asked:

(a) How often do you normally meet up with friends during your spare time?
(b) Who do you think understands you best, your parents or your friends?
(c) Who would you rather have fun with, your parents or your friends?
(d) Who would you rather be with during the evenings and at weekends, your parents or your friends?

All product moment correlation coefficients presented are significant at the <.001 level unless marked with an asterisk (*) in which case they are significant at the <.01 level. It should be observed that, since the analyses are conducted on a large sample ($n=1,334$), even small correlation coefficients (e.g., under .10) may be fully significant at the <.001 level. While, naturally, the conclusions drawn from them should be cautious, such correlations should in no way be discarded or ignored. Studies based on large representative samples are one of the major goals of the research enterprise, and it can be argued that their results often provide more accurate pictures of empirical reality than studies that are based on small, comparatively homogeneous, samples.

FINDINGS

Background Variables

In this section the relationships between the background variables of socioeconomic background and gender will be examined in relation to school achievement, school commitment, self-esteem, and VCR uses and preferences.

Socioeconomic background was measured by reference to father's occupational status and mother's education. These variables were themselves positively correlated, though to a lesser extent that might have been anticipated (.28). The analysis showed that the correlations between mothers' education on the one hand, and the school, self-esteem, and VCR variables on the other, were extremely weak and in most cases not significant. However, fathers' occupation revealed a number of significant correlations.

As expected, moderately strong positive correlations were found between fathers' occupation and school achievement (.27, .26, .27, in the 7th, 8th and 9th years respectively). Pupils from higher SES backgrounds also feel happier at school (.09) and are less eager to leave and start work (-.12).

Fathers' occupation was also weakly related to a number of the items measuring self-esteem and feelings about life. Adolescents from higher SES backgrounds are slightly more likely to feel that they have a lot of good qualities, that they are able to do things as well as their friends, that they are satisfied with their gender, and that there is a lot in life to be happy about. However, they are more likely to do things with friends that their parents don't approve of (.12), and, if there is a collision, tend more often to choose to do things with friends rather than parents. Adolescents from lower SES backgrounds give up more easily when things get difficult (.11); tend more to feel that teachers often make them feel good for nothing (.10); and feel that they often get blamed for things they haven't done (.07*).

In terms of VCR uses and preferences, adolescents from lower SES backgrounds watch more both on weekdays and at weekends (-.14,

-.12); and more often watch detective/police stories, karate, and films with a lot of explicit violence (.13, .16, .10*, respectively).

Females tend to do better at school, a correlation which increases slightly from the 7th to the 9th year (.17, .18, .19). They also tend to grade their schools higher; to do more homework; to be happier at school; and to be more positive about returning there after holidays (.17, .19, .11, .12). Conversely, males are slightly more likely to be longing to leave and start work (.06*).

In terms of self-esteem, the results showed consistent gender difference. Males are more satisfied with themselves, more often think that they have a lot of good qualities, feel more that they are able to do things as well as their friends, and are more satisfied with their looks (.16, .27, .12, .24). Conversely, females more often feel completely worthless, wish that they liked themselves more, more easily feel a failure, and wish that they could change themselves (.27, .21, .18, .18).

In feelings about life in general, however, gender differences were less consistent. Males more often feel that they get blamed unfairly for things they haven't done; that their parents often disapprove of their friends; and that people don't really care about what happens to others (.12, .10, .06*). Females slightly more often feel lonely and exposed, and that everything feels uncertain and insecure (.07*, .06*). There were no significant relationships between gender and the items: "There is a lot to be happy about here in life"; "Everything in life is getting worse"; "I often feel that life is meaningless"; and "I often feel bored."

The VCR use data show that males use VCRs more, both on weekdays (.15) and at weekends (.17). They also more frequently watch violence, karate, and pornography (.22, .26, .36,). However, gender was not significantly related to viewing horror films.

In summary, the results presented above show that both gender and father's occupation are significantly related to the central variables of this study. Taken together they suggest that both lower school achievement and greater VCR use, especially of violent contents, are characteristic of males in general, and males from lower SES backgrounds in particular (the masculine taste for pornography appears to be unrelated to fathers' occupational status). Moreover, there is some indication that the members of this group feel victimized in the sense of being blamed unjustly for things they haven't done. On the other hand, despite their higher school achievement, low self-esteem seems most characteristic of females in general; and, at least in some respects, of females from lower SES backgrounds in particular.

It would appear, then, that the relationships between school achievement, self-esteem and VCR use are not straightforward. Females have a higher average level of achievement but less positive feelings about themselves, whereas males have lower average levels of achievement but more positive self feelings. Furthermore, it is males, with their greater self-esteem, who have higher average levels of VCR use and more frequently watch violent films. Nevertheless, as we shall see, there is an overall negative correlation between school achievement and VCR uses and preferences. There are also correlations between the VCR variables and some aspects of low self-esteem. These results can be interpreted as supporting the finding, cited earlier, that academically successful females have lower opinions of themselves than do comparable male groups, whereas unsuccesful females react to their failure in more or less the same ways as do their male counterparts. This would account for the apparent incongruity of these results, while leaving the basic postulates of the model intact.

Peer Group Activity and Parent-Peer Orientation

As has already been noted, much adolescent use of VCRs in Sweden takes place in groups. The possibility that such use may also develop subcultural characteristics was also discussed. Before continuing, therefore, peer group activity and parent-peer orientation will be analyzed in relation to school achievement, self-esteem, and VCR use.

The frequency with which adolescents met up with friends during their spare time was negatively correlated with earlier school achievement (-.16, -.19, and -.18) indicating that lower achieving pupils see their friends more often. This negative relationship between school and peer group activity was also reflected in attitudes to school. Those that spend more time with friends are less happy at, and less positive toward school in almost every respect. Above all, they spend less time doing homework, think that most lessons are a complete waste of time, and are longing to leave and start work (.20, .14, and .15). They also have a tendency to think that teachers often make them feel good for nothing; and that they often get blamed unjustly for things they haven't done (.08, .11). However, they are slightly more likely to be satisfied with their looks (.08*).

Males who meet up more frequently with friends have a slight tendency less easily to feel themselves to be failures; while females who meet up more often with friends tend to feel less alone and exposed. This supports the view that peer group involvment functions (in different ways according to gender) to provide support for those who fail to achieve positive status and identity in other fields of social activity.

Significant correlations were obtained between peer group activity and amount and type of VCR use. Those who meet up more frequently with friends watch more video per week (.21) and also spend more time watching violence, karate, horror, and pornography (.26, .21, .22, .19, respectively).

A similar pattern of relationships was revealed in terms of parent-peer orientation. In addition, those who feel that they are understood better by their friends than by their parents and who prefer to spend their leisure time with peers rather than parents are more likely to feel bored, to think that everything in life is getting worse, that people don't really care about each other, and that life is meaningless, uncertain, and insecure.

It appears that the adolescent peer group provides a defence against unsatisfactory conditions both in the family and at school. Furthermore, it is peer activity, rather than merely an orientation away from parents and toward peers, which is more effective in this respect. This can be seen from the fact that those who are merely oriented to peers tend more to feel bored, pessimistic, cynical, and insecure, while those who are more active in their peer contacts do not. This suggests that it is alienation from parents and school in the absence of active peer support that creates these negative existential feelings.

School Achievement and School Commitment

While school achievement was negatively correlated with peer group activity, it was not significantly related to parent-peer orientation. However, as might be expected, achievement was related to pupils' attitudes to school. For example, rather strong correlations were obtained between earlier achievement and the extent to which one is happy at school (.34, .38, .39), and longing to leave school and start work (-.28, -.29, -.32). Moreover, pupils' grading of their school was found increasingly to correlate with the grades they themselves receive (.29, .34, .35).

As predicted, school achievement was significantly related to self-esteem and feelings about life. Low achievers are slightly less likely to be satisfied with themselves and tend to wish that they were different. They are also less likely to feel that they have a lot of good qualities, that they are able to do things as well as their friends, and that there is a lot in life to be happy about. Moreover, they are much more likely to feel that teachers often make them feel good for nothing (.32 in the 9th class), that they are unjustly blamed for things they haven't done (.21), and that their parents often disapprove of their friends (.14). They give up more easily when things get difficult (.25), feel that life is meaningless (.12) and is getting worse rather than better (.11).

Substantial support was also provided for the hypothesized relationship between school achievement and VCR use. In terms of time spent viewing, low achievers watch more video both on weekdays (-.28, -.29, -.30) and on weekends (-.25, -.25, -.26). School achievement is also clearly related to what adolescents choose to watch with the help of VCRs. Thus, low achievers watch more violence (-.25, -.26, -.28), karate (-.25, -.26, -.27), horror (-.20, -.22, -.23), and pornography (-.20, -.20, -.22). It should be noted that, not only were these correlations already established in the 7th school year, they also show a tendency to increase slightly over time.

In general, the pattern of results for school commitment reflects that for school achievement, although with weaker correlations. Similarly, amount of homework done was negatively correlated with amount of VCR use on weekdays, and with watching violence, karate, and pornography.

Self-Esteem and VCR Use

In general the correlation coefficients between VCR use and the indicators of self-esteem and feelings about life were rather weak. The strongest correlations were those with, "I often get blamed unjustly for things I haven't done," and "My parents often disapprove of my friends" (in each case .13). Next came, "Teachers often make me feel good for nothing" (.10), followed by, "people don't really care about each other" (.10), and "I am satsified with my looks" (.10). Conversely, those who do not feel that "there is a lot in life to be happy about" watch slightly more on weekdays (.07*). However, feeling worthless and a failure correlated with slightly less VCR use at weekends (0.7*, .08*). Although these correlations are weak and not unequivocal, they do seem to indicate some relationship between amount of VCR use and more negative feelings about self and life.

The relationships between these variables and types of content used provide a clearer picture. The indicator most clearly related to use of "undesirable" contents was, "I often get unjustly blamed for things I haven't done," which correlated with more viewing of violence (.20), karate (.12), horror (.11), and pornography (.21). This was followed by, "Teachers often make me feel good for nothing" (.15, .12, .12, and .17); and, "My parents often disapprove of my friends" (.12, .09*, .07*, .19). No other indicator correlated with all four content types.

"People don't really care about each other" correlated with more frequent use of violence, horror, and pornography (.10, .08*, and .13), while more frequent viewing of violence and pornography also correlated with "often feeling bored" (.10, .10), and "life is meaningless" (.10, .09*). Feeling insecure and uncertain about everything was related to more frequent use of pornography (.08*), while not feeling that there is a lot to be happy about in life was related to more frequent use of horror (.07*).

In some cases preferences were related to more positive self feelings. "I am satisfied with my looks" was associated with greater use of violence and pornography (.09, .15). "On the whole I am satisfied with myself" was associated with viewing more karate and pornography (.08*, and .12), as was "I think that I have a lot of good qualities" (.11, .12). Conversely, "sometimes I feel completely worthless" was associated with less viewing of karate and pornography (-.09*, -.12).

If we examine these relationships from the perspective of what is viewed, we see that the content category that manifests the greatest number of significant correlations with the self variables is pornography, followed by violence, and then karate and horror. It would, perhaps, not be stretching the point too far to say that this also reflects the order of the social disapprobation that is attached to these types of content.

Discussion

The basic postulation discussed and tested in this article is that, in the context of the highly charged moral debate surrounding the arrival of VCRs in Sweden, the uses made by Swedish adolescents of "undesirable" prerecorded video cassettes is at least partly determined by the need to sustain self-esteem in the face of denigration resulting from failure or other negative experiences at school. The results provide substantial support for this postulation, indicating that school achievement is quite strongly related to self-esteem, feelings about life, and VCR uses and preferences. Some support was also provided for the postulated relationship between self-esteem and VCR uses, although here correlations were weaker and less consistent. There are significant gender differences in these relationships which must be taken heed of but, overall, low achievers have lower self-esteem, more negative and pessimistic feelings about life, a higher level of VCR use, and more frequent viewing of pornography, violence, and other socially disapproved of contents.

These results can be located in a research tradition which sees school experience as a vital determinant of media and other cultural orientations. As long ago as in 1964, Stinchcombe pointed out that among those for whom the school holds out little promise of future status rewards, other cultural elements may be used as symbols of identity. It is these symbols that constitute much of youth culture. Where legitimate achievement systems fail to provide sufficient self-esteem, students substitute ascriptive symbols of adulthood, such as participation in otherwise forbidden activities.

In adolescence, the school and the peer group are central providers of the symbols of personal worth. Our results indicate that there is an inverse relationship between school achievement and frequency of peer activity, i.e., students with lower grades meet up with friends more often than do high achievers. We interpret this as indicating that some students, in order to reduce the punishment to their self-esteem associated with failure, switch their loyalty away from the school and make alternative investments in the peer group and its activities. It is this membership of valued groups and subcultures that enables adolescents to reconstruct their identity. In some cases this process involves the more or less conspicuous consumption of the "forbidden." Such consumption makes possible the public display of private qualities, enabling us to express our "selves" by means of our public presence (see also Frith & Horne, 1988).

This interpretation is supported by the fact that those who are more active with peers devote more time to VCR use and also spend more time watching violence, horror, and pornography. Significantly, they are also less bored, pessimistic, cynical, and insecure than adolescents who are oriented more to their peers than to their parents, but who meet up with friends less often. This implies that the peer group functions in a positive way as an alternative source of identity for those alienated from the school and/or their parents. The problem for society is that this may involve engagement in various deviant and delinquent activities.

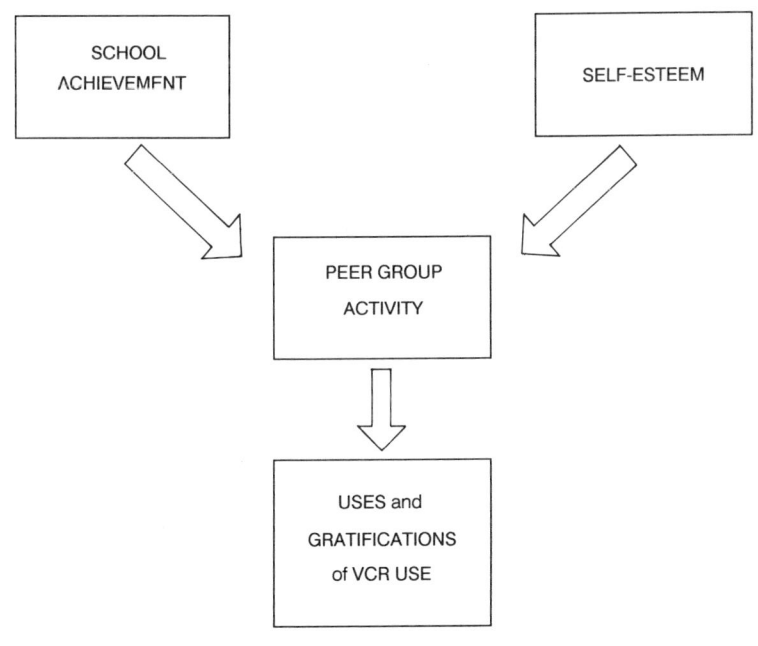

Figure 10.1. A Respecified Perspective on the Relationship Between School Achievement, Self-Esteem, and VCR Use

We conclude that the role of the peer group is the most important mediating factor in the set of relationships being addressed here. This also helps to account for the weak and partially inconsistent results obtained in the relationship between self-esteem and VCR use. Initially, a direct relationship was postulated between self-esteem and VCR use, i.e., it was hypothesised that VCR use of socially disapproved-of contents is stimulated directly by the need to defend and enhance self-esteem. We now see that this relationship is indirect; mediated by a peer group involvment that may, itself, successfully defend and enhance self-esteem and provide more positive feelings about life. In diagrammatic terms, this revised "model" may be represented as in Figure 10.1.

It follows that, essentially, the peer group *itself* is the first line of defence against threats to self-esteem, not what it spends its time doing.

Many activities are available to groups in respect of identity construction. These may be generally "positive," such as sports, music, and political activity or "negative," such as using a VCR for viewing brutal and explicit violence, vandalism, or delinquency. The choice will depend on three factors: the extent of alienation from parents and the school, the options available in the public and commercial provision of goods and services, and the reactions of the adult culture to the initial deviance.

With regard to the first two factors, the results have shown significant variation according to SES background and gender. Males, especially from low SES homes, tend to have lower achievement and more negative attitudes to school; they are also likely to have fewest resources in the market place of goods and services and tend to be overrepresented among groups labeled by society as deviant or delinquent. It is therefore hardly surprising that this group has a greater tendency to use VCRs in the ways indicated. This also accords with the view that youth subcultures are essentially celebrations of masculinity (Brake, 1980). For females the picture is more complicated. Despite an overall higher average achievement at school they have a lower average level of self-esteem, yet use VCRs less. Because of differential socialization and the status and role expectations of women in society, it is possible that male peer groups and subcultures provide better opportunities, via socially disapproved of activities, to resist, repair, and enhance self-esteem.

With regard to the third factor, it is possible that adults should show a little more circumspection in their reactions to adolescent identity construction. We need to heed Erikson's (1959, 1968) warnings not always to take adolescents' daring experimentation too seriously. In most cases, he argues, these "I dare you," "I dare myself" experiments are not harmful—provided they are not prematurely responded to with fatal seriousness by overeager and neurotic adults. For then, the adolescent may become exactly what fearful adults expect him to be—and make a total job out of it!

Finally, this study illustrates the heuristic value of the uses and gratifications paradigm that seeks to connect the psychological needs and social circumstances that lead to media use. The results have shown that the school situation is involved in the generation of media-related needs in four of the five ways identified by Katz et al. (1974: 27). Thus,

1. The school situation produces tensions and conflicts, leading to pressure for their easement via media consumption.

2. The school situation offers impoverished real-life opportunities to satisfy certain needs (in this case, for positive self-esteem and identity), which are then directed to the media.
3. The school situation gives rise to certain values, the affirmation and reinforcement of which is facilitated by the consumption of media materials.
4. The social situation provides a field of expectations of familiarity with certain media materials, which must then be monitored in order to sustain membership of valued social groupings.

Criticism of the uses and gratifications approach for its individualistic bias is scarcely new (see also Roe, 1987b), and it is apparent from the results presented here that the links between social situation, need fulfilment, and media use need to be refined, at least in the case of adolescents, to include the essential mediating role of the peer group. Furthermore, this study illustrates how individual need fulfilment takes place in a group context firmly located in the social structure. Future research needs to address all three levels of analysis if it is to provide an adequate picture of adolescent media use.

REFERENCES

Atkins, C. K. (1972). Anticipated communication and mass media information-seeking. *Public Opinion Quarterly, 36*.

Bandura, A. (1978). The self system in reciprocal determination. *American Psychologist, 33*. 344-358.

Bourdieu, P. (1967). Systems of education and systems of thought. *International Social Science Journal, 19,* 3, 336-358.

Bourdieu, P. (1977). Cultural reproduction and social reproduction, in A. H. Halsey and J. Kaubel (Eds.), *Power and ideology in education.* New York: Oxford University Press.

Bourdieu, P. (1979). *The inheritors: French students and their relation to culture.* Chicago: Univ. of Chicago Press.

Bourdieu, P. (1984). *Distinction: A social critique of the judgement of taste.* London: Routledge & Kegan Paul.

Brake, M. (1980). *The sociology of youth culture and youth subcultures.* London: Routledge & Kegan Paul.

Cohen, A. K. (1970). A general theory of subcultures. In D. O. Arnold (Ed.), *The Sociology of Subcultures.* Berkeley, CA.

Cohen, S. (1980). Folk devils and moral panics. Oxford, UK: Martin Robertson.

Dembo, R. (1972). Life-style and media use among English working class youths. *Gazette, 18,* 24-36.

Dembo, R. (1973). Gratifications found in media by British teenage boys. *Journalism Quarterly, 50,* 517-526.
Ds U. (1987: 8). *Videovald. En rapport fran valdsskilldrings-utredningen.* Stockholm, Sweden: Utbildningsdepartmentet.
Elder, G. H. (1968). Adolescent socialization and development. In E. F. Borgatta & W. W. Lambert (Eds.), *Handbook of personality theory and research.* Chicago: Rand McNally.
Erikson, E. H. (1959), Identity and the life-cycle. *Psychological Issues. 1.*
Erikson, E. H. (1968), *Identity, youth and crisis.* New York: Norton.
Frith, S., & Horne H. (1987). *Art into pop.* London: Methuen.
Gahlin, A., & Nordström, B. (1988), Video i Sverige. *Sveriges radio: Publik- och programforsknings avd.* Report no. 6.
Greenberg, B. S., & Heeter, C. (1987). VCRs and young people. *American Behavioral Scientist, 30*(5), 509-521.
Gunter, B. and Levy, M. R. (1987), Social Contexts of video use. *American Behavioral Scientist, 30*(5), 486-494.
Hargreaves, D. H. (1967). *Social relations in a secondary school.* London: Routledge & Kegan Paul.
Hulten, O. (1985). Elektroniska medier. Aktuell utveckling i Sverige. *NORDICOM-Nytt Sverige.* Temanummer "Medier i utveckling." 1-17.
Höjerback, I. (1986). Video i Malmö. *Forskningsrapport i kommunikationssociologi* no. 1. Lund, Sweden: Dept. of Sociology.
Katz, E., & Foulkes, D. (1962). On the use of mass media for "escape": Clarification of a concept. *Public Opionion Quarterly, 26.*
Katz, E., Blumler, J. G., & Gurevitch, M. (1974). Utilization of mass communication by the individual. In J. G. Blumler & E. Katz (Eds.), *The use of mass communications.* Beverly Hills, CA: Sage.
Levy, M. (1987). Some problems of VCR research. *American Behavioral Scientist, 30* (5), 461-470
Mann, D. W. (1981). Age and differential predictability of delinquent behaviour. *Social Forces. 60*(1), 97-113.
Maslow, A. H. (1968). *Toward a psychology of being.* New York: Van Nostrand.
Murdock, G., & Phelps, G. (1973). *Mass media and the secondary school.* London: Macmillan.
Nordisk Medie Nyt. (1988). no. 3. p.9. Copenhagen, Denmark.
Power, C. (1986). The purdah experience. In P. Fensham (Ed.), *Alienation from schooling.* London: Routledge & Kegan Paul.
Purkey, W. W. (1970). *Self-concept and school achievement.* Englewood Cliffs, NJ: Prentice-Hall.
Roe, K. (1981). *Video and youth: New patterns of media use.* Media Panel Report no. 18. Lund, Sweden: Dept. of Sociology.
Roe, K. (1983a). *The influence of video technology in adolescence.* Media Panel Report no. 27. Lund, Sweden: Dept. of Sociology.
Roe, K. (1983b). *Mass media and adolescent schooling.* Stockholm: Almqvist & Wiksell.
Roe, K. (1985). The Swedish moral panic over video. *NORDICOM Review of Nordic Mass Communication Research, 1,.* 20-25.
Roe, K. (1987a). Schooling ourselves to heavy metal. In I. Frones & O. Stafseng (Eds.). *Ungdom mot ar 2000.* Oslo, Norway: Gyldendal.

Roe, K. (1987B) Adolescents video use: A structural-cultural approach. *American Behavioral Scientist*. *30* (5), 522-532.
Roe, K., & Johnsson-Smaragdi, U. (1987). The Swedish "mediascape" in the 1980's. *European Journal of Communication*, *2*, 357-370.
Roe, K., & Löfgren, M. (1981). Music video use and educational achievement: A Swedish study. *Popular Music*, *7*, 297-308.
Roe, K., & Salomonsson, K. (1983). *The uses and effects of video viewing among Swedish adolescents*. Media Panel Report no. 31. Lund, Sweden: Dept. of Sociology.
Rogers, C. R. (1951). *Client centered therapy*. Boston: Houghton Mifflin.
Rosenberg, M. (1965). *Society and the adolescent self-image*. Princeton: Princeton University Press.
Rosengren, K. E., & Windahl, S. (1972). Mass media consumption as a functional alternative. In D. McQuail (Ed.), *Sociology of mass communications*. Harmondsworth: Penguin.
Rosental, R., & Jacobson, L. (1968). *Pygmalion in the classroom*. New York: Rinehart and Winston.
Salomon, G. (1981). *Communication and education: Social and psychological interactions*. Beverly Hills, CA: Sage.
Schostak, J. (1986). Schooling the violent imagination. London: Routledge & Kegan Paul.
S.O.U. (1988: 28) *Videovald II—forslag till atgarder. Betankande av valdsskildringsutredningen*. Stockholm, Sweden: Allmanna forlaget.
Stinchcombe, A. L. (1964). *Rebellion in a high school*. Chicago: Quadrangle.
Sullivan, H. S. (1953). *The interpersonal theory of psychology*. New York: Norton.
Svendsen, K. H., & Vildsvik D. H. (1987). *Ungdom og video*. Bergen: Psykologiske fakultet.
Wall, J. (1988). *Skolbarns videotittande och videofilmer*. Stockholm, Sweden: I. M. S.
Willis, P. (1978). *Profane culture* London: Routledge & Kegan Paul
Wylie, R. C. (1968). The present status of self theory. In E. F. Borgatta and W. W. Lambert (Eds.), *Handbook of personality theory and research*. Chicago: Rand McNally.

PART IV
VCRs, GROUPS, AND SOCIETIES

11

Away from the Mainstream? VCRs and Ethnic Identity

JULIA R. DOBROW

The videocassette recorder was heralded by many as the technology that could liberate viewers from both the rigors of television scheduling and from the relative narrowness of television content by enabling people to watch what they want to watch when they want to watch it. Levy (1987) pointed out that while such flexibility could allow for, if not promote, greater cultural diversity in self-programming, to date this potential has gone largely unexamined.

This chapter examines whether different cultural subgroups use the VCR as a vehicle for what Gans (1974) called *narrowcasting*, the viewing of content that is more culturally or subculturally specific than what is usually shown on network and cable television in the United States. This work represents a preliminary foray into the area of finding and defining this issue. It addresses the ways that some groups of immigrants might use the VCR as a way to facilitate ethnic identity, create an ethnic boundary, or define *they* and *we*.

Most work in mass communications has focused on the identification and uses of messages distributed to a large and heterogeneous audience, while relatively little work has sought to understand the social dynamics of messages that reach smaller, more specialized, and possibly homogeneous audiences. Large audiences have been of interest for both economic and theoretical reasons. It has been postulated that the most prevalent distributor of messages in our culture, over-the-air television, disseminates a relatively stable set of images to large numbers of people, cultivating mainstream perspectives among otherwise divergent groups of viewers (Gerbner, 1987). But, because of their flexibility and versatility, VCRs enable viewers to use a television set for additional purposes. This technology thus raises other sets of questions: do subgroups of viewers use the VCR for purposes apart from viewing mainstream television broadcasts? Does VCR use diversify the viewing ritual, taking viewers away from the centralized set of coherent messages to individualized ones which might contribute to the cultivation of individual perspectives?

The present work focused on three research questions:

1. What kinds of programs did members of these ethnic communities use the VCR to watch?
2. Were VCRs being used as a substitute for or extension of other media?
3. What were the social conditions surrounding viewing?

The significance of investigating such issues is clear. If, indeed, the VCR enables subgroups to view other than mainstream programming, this could alter traditional notions of a mass audience. Research on VCR use in countries other than the United States has suggested that different ethnic and political subgroups use VCRs for purposes including the viewing of specific ethnic films, homemade documentations of family and community events, and the use of cassettes as a form of exchange (see, for example, Koh, 1982; and Suraiya, 1983). Will VCR use in this country follow similar patterns, now that the technology is penetrating ever-increasing segments of the American population? Since VCR penetration in 1988 constituted more than 50% of all American television households ("VCR usage on fast forward," 1988) and is projected at anywhere from 60%-90% by the end of the decade by different forecasters (*Video Marketing*, 7 October 1985; Brotz, Wyche & Trautman, 1986), it is logical to assume that VCRs are increasingly available to and used by diverse groups of people. The widespread use of VCRs could have far-reaching economic consequences for media industries which traditionally have aimed toward a mass audience, as well as far-reaching social consequences if VCRs enable the cultivation of individual perspectives, away from the mainstream.

THE ETHNIC BACKGROUND

Social scientists, historians, and novelists (among others) have long argued over the definition of ethnicity. The debate generally has focused on two issues: (1) whether ethnicity is determined by the categorization of nonmembers on the basis of sociocultural issues and is thus an ascribed characteristic or whether ethnicity is determined by the assumptions and sociocultural norms and values that a group makes about its own members, involving a component of voluntarism and is thus achieved; and (2) in defining ethnicity, is it operative *only* in comparative situations or is ethnicity a stable and constant state unique to a group, *regardless* of surroundings?

The position that ethnicity can be defined only in relationship is exemplified by A. Cohen (1974). In Cohen's view, ethnicity is only a conscious issue in urban or other settings in which different groups coexist, and some are more prevalent in number, economic, political, or social status than others. Epstein (1978) concurred with this position, and wrote that it is only meaningful to talk about ethnicity where groups are in interaction with one another within some common social context. Ronald Cohen (1978) stated that "ethnicity is first and foremost situational . . . the interactive situation is a major determinant of the level of inclusiveness employed in labeling self and others" (p. 338).

Other social scientists, however, have defined ethnicity as internally ascribed rather than externally labeled and contend that its presence is not limited to times of change or conflict. Barth's (1969) primary emphasis was categories of ascription and identification by group members, themselves. This departure from other definitions has two important implications: first, the definition makes no assumptions about the content of ethnicity, thus ethnicity is an ". . .organizational vehicle that may be given varying amounts and forms of content in different sociocultural systems and situations," and second, it implies that what gives a group its distinctive ethnicity is not so much its internal content as where its ethnic boundaries are drawn (p. 11). Following Barth's lead, Vincent (1974) wrote that "what ethnicity requires in its barest essence is 'we' and 'they' . . . but ideas of negation and 'otherness' are also important for the definition of the social unit and for the delineation, maintenance and transformation of boundaries" (p. 576).

Finally, there are some definitions of the term *ethnicity* that attempt to incorporate both internal and external labeling. DesPres (1975), for example, defines ethnicity as both an objective and subjective phenomenon. "Ethnic groups are defined both by the objective cultural modalities of their behavior . . . and by their subjective view of themselves and each other" (p. 192). Peterson-Royce (1982) suggested that a group's ethnicity is composed of "double boundaries," and wrote that "the boundary maintained from within and the boundary imposed from outside . . . result from the process of interaction with others" (p. 29). Rose (1981) has posited that members of various ethnic groups are apt to feel their ethnicity more intensely when it is " . . . determined by the attitude of members of the 'host' society toward the 'strangers in their midst.'" (p. 7).

Ethnic affiliation has been of interest to communications researchers for both theoretical and empirical reasons. McQuail (1984) has written that on an individual level, media might help to develop or to diffuse value systems, and on a larger societal level, might not only reflect society, but help to mold it. Applying these concepts to the issue of ethnic identity, Berry and Mitchell-Kernan (1982) suggested that the representation (or lack thereof) of ethnicity on television might well influence children's perceptions of their own and others' ethnicity, as well as play some part in their socialization and the development of their values about others.

Gerbner's cultivation theory (1987) posited that exposure to mass produced messages creates (among those who expose themselves to these messages) commonalities, rather than differences. This so-called "cultivation of dominant image patterns" might suggest that media use would blur lines of ethnic affiliation, as well as create media stereotypes that could affect individual and social values. All of these theories lend credence to the idea that the media might also play some role in determining when ethnic affiliation would be ascribed, and when it might be achieved.

While there is much literature which examines ethnic background as a demographic variable in media behavior, relatively little communications scholarship has directly addressed the relationship of ethnic affiliation and media use. There is a school of thought and some empirical evidence for an assimilationist theory, suggesting that exposure to the media of the dominant society (or what some have called mainstream media) encourages assimilation to its norms and values by members of nondominant groups (see, for example, Goldlust and Richmond, 1974; Kim, 1978; 1979; 1984). On the other hand, there is a school of thought and some empirical evidence for a more pluralistic perspective—studies done on the maintenance and use of native language media (mostly newspapers and radio) show that their use in combination with or to the exclusion of mainstream media can develop a "consciousness of kind," and might slow down or delay the process of assimilation among some groups of users (Breton, 1964; Jeffres & Hur, 1981). However, as Subrvi-Velez (1986) points out, distinctions between just what is called *assimilation* and *pluralism* are difficult to operationalize, and empirical work seeking to make such distinctions must be examined carefully.

Many researchers have suggested that variables at the level of individual differences do not explain much about exposure to television. For example, Bower's studies (1970; 1985) on the relationship between viewers' background characteristics and their viewing behavior found few significant differences. He concluded that the television audience was ultimately "extraordinarily undifferentiated" by any sociocultural characteristic. Other researchers, including Gerbner (1987), insist that "most viewers watch by the clock" and are differentiated from one another only by the amount of time they spend viewing.

On the other hand, Gans (1974, p. 133) suggested that it was certainly possible for viewing to be defined by social group. He postulated that "subcultural programming" (the individualized use of media by members of different "taste publics" defined by class, age, race and other social factors) would ". . . enable audiences to find content best suited to their wants and needs, thus increasing their aesthetic and other standards, and the relevance of their culture to their lives." Wilson and Gutierrez (1985), for example, pointed out that members of ethnic and racial minority groups often make use of specialized minority programming in the few instances where it is available (though this is primarily in terms of specialized radio stations and newspapers rather than television programming).

VCRs in particular do provide an opportunity—at least in principle—for the type of specialized programming through repackaging particular types of imported movies or television shows on videocassette that makes individualized programming economically feasible and might well promote social cohesion through media use.

But who is the VCR public? Shortly after the VCR's introduction in the United States in 1976, David Lachenbruch wrote that the 40,000-odd Betamax owners ". . . probably became the most questionnaired and market-researched segment of the American public" (1977).

The demographics of the early VCR market were predictable and quickly verified. Most studies of the U.S. market found that VCR owners were a well-educated and economically upscale audience (Kalba-Bowen Associates, cited in Agostino, Terry, and Johnson, 1980; Levy, 1980; 1983). However, more recent studies have indicated that while the majority of the VCR owning public continues to be consistent with the early profiles, the declining cost of VCRs and the increasing availability of rental units has brought some diversification (*Media Matters*, July 1986).

It is clear that in countries other than the United States, VCRs have penetrated more quickly throughout the population, and that VCR owner profiles may be more differentiated that they have been among American early adopters (Boyd and Straubhaar, 1985; Straubhaar and Lin, 1986). In Malaysia, for example, where government officials are trying to unite the three major ethnic groups through encouraging use of the national Bahasa Malay language in the media, the growth of imported VCRs and Chinese language tapes from Singapore and Hong Kong pose questions of assimilation, acculturation and politics (Koh, 1982). Ogan (1988) postulated that as VCR ownership in many countries increases, there are increased opportunities to avoid centrally distributed broadcasts and increased opportunities to view individualized content.

We are starting to see similar patterns in the United States, though to date no large-scale study has documented this. Dobrow (1984) noted that VCRs were being used by groups of Indian immigrants in the greater Philadelphia area, people who did not conform to the identified characteristics of the majority of U.S. VCR users. Snyder (1985) reported many groups of immigrants in the Boston area using VCRs to view videocassettes in their native languages. These scattered studies are significant for two reasons: first, because they raise the question of what and who might really constitute a "mass audience" in the late 1980s. As Williams et al. (1985) wrote, ". . .the concept of 'audience' itself becomes problematic. . . . Perhaps many assumptions about audiences need to be re-examined" (p. 251). And second, if there are, indeed, distinguishable ethnic groups of viewers, perhaps their use of the technology can begin to tell us something about the ways in which VCRs might be used as a way of maintaining ethnic identity.

METHODOLOGY

The present study is part of a larger study of video use patterns conducted in the spring of 1986. Sixty-two in-person interviews were conducted with individuals who identified themselves as immigrants or as the children of immigrants to the United States. The interviews took place in seven video stores, and in four stores which sell speciality food or clothing and also rent tapes. (Earlier work [Dobrow, 1984] and pretesting had revealed that several ethnic food stores also rent videocassette tapes of non-American movies and television programs.) The stores that comprised the sample were located in demographically diverse parts of the greater Boston area, including stores in suburban and more urban areas. These stores were identified primarily from tips by respondents. Stores visited included one that rented Japanese and Filipino tapes, one that rented Russian tapes, one that rented Indian tapes, and one that rented Greek and Arabic tapes. All stores that permitted the author to conduct research on their premises were included as research sites. Interviews took between 10 and 25 minutes to complete. Each of these interviews was taped and later transcribed.

The sample which will be discussed in this chapter was limited to those interviews conducted with members of immigrant ethnic groups, defined here as those people who willingly identified themselves as having come to the United States from another country or who noted that their immediate relatives had done so. Some, who continued to identify themselves by their particular ethnic background, represented the third generation removed from immigration. Respondents came from the following countries: Egypt (n = 12), Greece (n = 16), India (n = 9), Japan (n = 12), the Philippines (n = 4), and the Soviet Union (n = 9).

In all, 62 interviews were conducted with these VCR owners or renters. Fifty-two percent were female, and 48% male. The majority of this sample were between age 36-64, though respondents ranged in age from 18 to over 70. They had diverse educational and economic backgrounds. While some of these informants had had their VCRs for less than a year (12%), the majority had owned or rented machines for two or more years.

RESULTS

In response to the first research question, for what types of programming did this sample use their VCRs, virtually every informant reported using his or her VCR primarily for the viewing of foreign language videocassettes and, to a lesser extent, for the viewing of rented American films. Most of the cassettes that the informants mentioned were movies from their country of origin, though some were taped broadcasts of foreign television shows, sporting events coming from a country other than the United States, or even home-made videos sent by relatives or friends in the home country.

Few respondents reported using the VCR to time-shift television programs. The tapers were almost entirely parents, several of whom said that they taped programs such as "Sesame Street," but not only for the benefit of their children. One mother explained, "This program I watch with my children because it helps me, too, with my English!"

Almost all of the first generation immigrants who were interviewed, regardless of national background or of how long they had been in the United States, had similar explanations for their rental behavior. "These Russian movies, they make me feel connected to the old country" said an elderly Russian man who has been living here for four years. "We know these actors, we grew up with them. It is a way of bringing us back home—we can forget the bad parts that are in them [the propagandistic elements] and remember the good, the core of Russia, the part that is still with us."

His comments highlighted another recurring theme: in viewing tapes on the VCR, immigrants were able to selectively perceive their own backgrounds. As an Indian woman said, "We left [India] because things were very bad. But when we watch the movies, we only see what was good." A young Egyptian man explained, "I left Egypt to get a better job in America—it was a very depressed economy at home. But when I came here, I liked to watch the tapes [movies in Arabic] . . . they showed me the place I was homesick for."

Many immigrants reported viewing foreign language video tapes because the movies or programs brought them in touch with their first homes. In these cases, video tapes appeared to enable the immigrant to define his or her ethnicity in the midst of assimilating forces. A woman in her mid-60s who has spent all of her adult life in America still rents

Greek tapes about once or twice a month. "I grew up there, and I like to see these things. They remind me of my childhood, of my parents, of where my roots are."

A Filipino woman reported that her relatives in Manila regularly sent her "progress reports" on family and friends via video. "I have young nieces and nephews growing up there," she explained. "This way I can see how they're changing. I can keep in touch, and I don't feel so far away from home."

A young Egyptian-American recalled that when he first came to the United States, more than six years ago, he would drive 50 miles to see an Egyptian movie. "I was desperate to see things that reminded me of home. These films were my only tie. There were a group of us who drove from Boston to Lawrence and back, every week or so, just to see Egyptian movies!"

His comments were representative of many immigrants and shed some light on the third research question about the social conditions under which the respondents used their VCRs. Many respondents reported gathering in groups to view the different native language videotapes. The groups often contained other than just family members. A Japanese woman said that she and several other women, all of whom had married American soldiers, gather together and have tea parties each week. An important feature of these tea parties is the viewing of a Japanese film or television program on cassette: "We watch [them] all together and talk about them. No one else here knows our experience as we do. We talk in Japanese and talk about home." An Indian man explained, "There is a small but tightly knit [Indian] community here. We regularly gather to watch the films together. We eat Indian foods. We wear traditional Indian clothes, which we don't always wear. It reminds us all of home, and of how we are Indians in another land."

Some respondents, particularly first generation immigrants, said that they prefered to watch programs on the VCR to the exclusion of watching American television programs. "There is so much bad stuff, so much violence, so much sex [on American shows]" said one Russian woman. "I cannot watch it. But with the VCR, I can watch what I want to. I need not see anything else." For others, however, particularly those who were younger than age 30, the VCR appeared to be an extension of other media; several people stated that they used it to "get some variety from what is always on television in America." Within this subsample, there was no apparent relationship between the length of time one's family had been in the United States and the extent to which the primary VCR use was to view foreign language cassettes.

Not surprisingly, the children of first generation immigrants, from their own accounts as well as from their parents', were less attracted to viewing foreign language videos. The twenty-five year old son of Greek parents said "My parents like me to watch these Greek movies with them, as though some of the culture or something would seep in. But I don't know the language—literally, it's all Greek to me, and I'm not as interested in learning it as they [parents] would like me to be." The teenage daughter of a Japanese couple said, "My mom likes me to watch her Japanese movies with her, and she tries to teach me stuff about the language. I guess that stuff is OK, but I'd rather watch American movies." One of the Filipino-American respondents said, "At first, when we came here, I'd watch the Filipino movies with my family. But now, I don't know if I'm becoming more American or just getting older, but I don't like them as much anymore. I'd rather watch old American films on cassette."

The parental point-of-view was summed up by a Russian father who said:

> "It is hard, yes, when children want to be Americans, and forget that their parents were not born here. But it is important to me that they learn about Russian themes. And the Russian movies, they are not as violent as the American ones, with all the guns, guns, guns. I want my children to see Russian movies because the Russians in American movies are evil, are drunks, and bad. How many Russians are like that? Children watching the American movies would think all Russians are evil and drunks. I want my children to be Americans, but I want them to be proud of their heritage."

The cycle of immigration and connection to viewing foreign language videos became completed by interviews with some third generation "hyphenated-Americans". (As Rose [1981] explained, this phrase, emanating from John Dewey, refers to the trend that began in the early twentieth century to label one's self as a mixture of ethnicity of origin and immigrated nationality, as in Italian-American, Irish-American, etc.). These third generation respondents reported that they liked to rent videocassettes of their grandparents' countries of origin to learn about the language and culture. "I take Russian classes in college," said one young man. "My grandparents came from Russia. They're not around anymore, and my parents don't know anything. But I want to find out about my past. I watch some of these Russian movies, not that I can understand everything that goes on, but I try to get a sense of the culture, the values." A twenty-one year old said that he regularly rents Japanese movies to help him learn the language. "It's been a long time since anyone in my family really had anything to do with Japan, but I'm really curious. I'm trying to learn the language, because I'd like to go visit, and this is a good way to do it."

DISCUSSION

It is clear that some immigrants to the United States and their families use their VCRs as a way to view mass media and other content out of the mainstream of American television programming. Moreover, nonimmigrant members of the ethnic groups studied also used the VCR to see non-American programming.

In a sense, those who reported using the VCR for purposes like those described here can perhaps better be seen as taste publics (Gans, 1974; Bourdieu, 1984), rather than as members of a mass audience. People from many different demographic backgrounds intimated that they watched programs on the VCR because they could select the material they wanted to see and control when they wanted to see it. The significance of these data, as well as of this framework for analyzing them, is that both speak to a different formulation of the audience for mass media, one defined not only by the usual socioeconomic demographic characteristics, but also a large number of social and cultural factors that lead people to make similar sorts of decisions, ones that may not be a part of the mainstream.

As much of these data have demonstrated, people seem to be using VCRs, in effect, to do their own programming. If, indeed, this kind of narrowcasting along the lines of ethnic affiliation were ever to become more widespread, the economic and political ramifications for American broadcasting could be far-reaching. Narrowcasting and subcultural programming could mean a restructuring of the priorities of the existing network, cable, and pay television industries. Programming decisions and advertising strategies would have to be refined to address markets stratified by ethnic affiliation. Access to information and common cultural codes might be irrevokably altered.

Moreover, these data have shown that at least for the respondents interviewed, viewing of programs on the VCR often provide a forum for social interaction. We sometimes think of television viewing as being a solitary activity, viewed most by lonely or isolated people. But the indications from the present research are that VCR viewing might be quite different. These results contradict those of Gunter and Levy (1987), who found among their sample individualized and privatized use of this media. It may be that among members of an ethnic minority group, viewing becomes a communal activity precisely *because* members use it as a forum for defining or reinforcing ethnic affiliation, differences between themselves and the host society. Other researchers have reported video viewing in groups as a possible mechanism for achieving social solidarity (Roe, 1987; Ogan, 1988). It is possible that these cassettes from other countries could provide a forum for identity and solidarity in a way mainstream American network television broadcasts could not.

Rose (1981) wrote that members of ethnic groups are most likely to feel their backgrounds most intensely when expression of its cultural elements is distinctly different from that of the dominant society's culture. VCRs give ethnic group members the ability to experience their differences from the mainstream in a private setting, but, ironically, through a mainstream medium. It is possible that viewing ethnic videocassettes could supplement, or even replace, more public forms of ethnic affirmation. Many other social and cultural changes in our society could occur if privatization of ethnic expression becomes a by-product of VCR use.

In addition, if the viewing of individualized ethnically oriented video tapes by members of cultural subgroups were ever to become widespread, there is another cultural consequence to consider. As Wilson and Gutierrez (1985) point out, "Audience segmentation can also mean that minorities become further separated and, possibly, distanced from the rest of society. Segmentation points to a society in which people may be integrated in terms of the products they consume, but do not share a common culture based on the content of the entertainment or news media they use" (p. 233). In this sense, media use might well help serve to ascribe ethnic boundaries between self and others.

Finally, there was some indication that VCR use by ethnic subgroups might be cyclical. As the old adage suggests, what the son wants to forget, the grandson wants to remember. Some children of immigrants reported not wanting to watch the same foreign language videocassettes that their parents wanted them to see—and some second or third generation members of ethnic minority groups reported wanting to rent videocassettes to learn about the culture of their grandparents or parents. In this way, a study of how people are using VCRs can illuminate other social and cultural patterns, and is surely something worthy of future investigation.

Will VCR use enable members of ethnic minority and immigrant groups to expose themselves to other than mainstream American programming and begin to cultivate their individual perspectives? The data presented here give some indication that the technology can be—and *is* being—used to view mediated content selectively and to begin to narrowcast. But the extent to which this will form the framework for the cultivation of pluralistic perspectives is a question which must be explored when VCRs have penetrated even further into the American population.

In addition, this study suggests five major areas for future investigation:

1. *Method*

    Work on subgroups of users might well be difficult to study with traditional survey research techniques alone, since they are often difficult to locate in sufficient numbers with most sampling methods and may have language and/or cultural differences rendering telephone or paper-and-pencil questionnaires inadequate. However, they might be better studied utilizing multimethod approaches incorporating ethnographic data. Clearly, this work needs to be done with larger and more diverse samples. Longitudinal studies would be helpful in determining how, if at all, VCR use varies among individuals at different stages of the pluralism-assimilation continuum.

2. *Comparative research*

    Variables which need to be examined in greater depth include how an immigrant's length of stay in a host country affects VCR use, whether there are discernable differences between different ethnic groups in video use, and whether patterns can be traced cross-culturally. This study only examined immigrants from six nations: are use patterns of immigrants from other countries significantly different? Is it possible that immigrant use patterns relate more to video use patterns in their countries of origin than to their status as immigrants?

3. *Videocassette distribution*

    As VCR use becomes even more widespread, will we continue to find foreign language cassettes distributed primarily through ethnic food stores or informal trade within a community or will some of the mainstream video stores begin to carry them? If so, what would be the effects on the economics of the U.S video industry?

4. *VCR use and other media use*
   It might be enlightening to examine the extent to which the viewing of individualized videocassettes displaces other media use. Under what conditions will members of ethnic groups view mainstream American videocassettes or time-shifted American television programming, and when do they view foreign language cassettes? Do these patterns change over time?

5. *Social context of viewing*
   The data presented in this chapter show that among these subgroups, viewing usually occurs in groups. The extent to which the structure of the immigrant family group and community affects media use and VCR viewing patterns is a potentially rich area for future research.

The data presented in this chapter also indicated that the viewing of ethnic video material was often associated with other rituals (dress, food, etc.) Will this type of viewing continue as more people own VCRs or is this behavior an artifact of VCR distribution? What is the social meaning participants attach to the behavior of social rituals connected to video viewing? Or will it disappear entirely when assimilation takes place?

Ultimately, the use of VCRs by members of different ethnic subgroups poses an even more complex issue. Will the VCR unify diverse groups by providing increased exposure to common cultural content or will VCRs expand individualized viewing and narrowcasting, create multiple viewing centers, and take viewers away from the mainstream?

REFERENCES

Agostino, D., Terry, H., & Johnson, R. (1980). Home video recorders: Rights and ratings. *Journal of Communication 30*, 28-35.

Barth, F. (1969). *Ethnic groups and boundaries: The social organization of cultural difference.* Boston: Little, Brown.

Berry, G., & Mitchell-Kernan, C. (1982). *Television and the socialization of the minority child.* New York: Academic Press.

Bortz, P., Wychem, M., & Trautman, J. (1986). *Great expectations: A television manager's guide to the future.* Washington, DC: National Association of Broadcasters.

Bourdieu, P. (1984). *Distinction: A social critique of the judgement of taste.* Cambridge, MA: Harvard University Press.

Bower, R. (1985). *The changing television audience in America.* New York: Columbia University Press.

Bower, R. (1970). *Television and the public.* New York: Holt, Rinehart and Winston.

Boyd, D., & Straubhaar, J. (1985). Developmental impact of the home video recorder in third world countries. *Journal of Broadcasting and Electronic Media* 29(1).
Breton, R. (1964). Institutional completeness of ethnic communities and the personal relations of immigrants. *American Journal of Sociology* 70, 193-205.
Cohen, A. (1974). *Urban ethnicity.* New York: Tavistock.
Cohen, R. (1978). Ethnicity: Problem and focus in anthropology. *Annual Review of Anthropology (7)*, 379-403.
DesPres, L. (1975). *Ethnicity and resource competition in plural societies.* The Hague: Mouton.
Dobrow, J. (1984). *Ethnicity and identity: Television as an ethnic boundary.* Unpublished paper. Philadelphia, PA.
Dobrow, J. R. (1987). *The Social and cultural implications of VCR use: How VCR use concentrates and diversifies viewing.* Unpublished doctoral dissertation, Philadelphia, PA.
Epstein, A. (1978). *Ethos and identity.* Chicago: Aldine.
Gans, H. (1974). *Popular culture and high culture.* New York: Basic Books.
Gerbner, G. (1987). Mass media discourse: Message system analysis as a component of cultural indicators. In J. Bryant & D. Zillman, (Eds.), *Perspectives on Media Effects.* Hillsdale, NJ: Erlbaum.
Gerbner, G., Gross, L., Morgan, M., & Signorielli, N. (1980). The "mainstreaming" of America. *Journal of Communication, 30*(3), 10-29.
Glazer, N., & Moynihan, D. (1975). *Ethnicity: Theory and experience.* Cambridge, MA: Harvard University Press.
Goldlust, J., & Richmond, A. H. (1974). A multivariate model of immigrant adaptation. *International Migration Review 8*(2), 193-225.
Gunter, B., & Levy, M. (1987). Social contexts of video use. *American Behavioral Scientist, 30*(5), 486-494.
Isaacs, H. (1975). Basic group identity: The idols of the tribe. In Glazer, N. & Moynihan, D. (Eds.), *Ethnicity: Theory and experience.* Cambridge, MA: Harvard University Press.
Jeffres, L. W., & Hur, K. K. (1981). Communication channels within ethnic groups. *International Journal of Intercultural Relations 5*, 115-132.
Kim, Y. Y. (1978). Acculturation and patterns of interpersonal communication relationships: A study of Japanese, Mexican and Korean communities in the Chicago area. In Subverbi-Velez, F. *Communication Research 13*(1), 71-96.
Kim, Y. Y. (1979). Toward an interactive theory of communication-acculturation. In Nimmo, D. (Ed.), *Communication yearbook 3.* New Brunswick, NJ: Transaction.
Kim, Y. Y. (1984). Searching for creative integration. In Gudykunst, W. B. & Kim, Y. Y. (Eds.)., *Methods for intercultural communication research.* Newbury Park, CA: Sage.
Koh, F. (1982). A dilemma in view. *Far Eastern Economic Review 117*, 46-48.
Lachenbruch, D. (1977). The new boom in VCRs. *House and Garden 149*(1): 62-63.
Levy, M. (1980). Program playback preferences in VCR households. *Journal of Broadcasting and Electronic Media 24*, 327-336.
Levy, M. (1983). The time-shifting use of home video recorders. *Journal of Broadcasting and Electronic Media 27*, 263-268.
Levy, M. (1987). Some problems of VCR research. In Levy, M. (Ed.), *American Behavioral Scientist 30*(5): 461-470.
McQuail, D. (1984). *Mass communication theory.* Beverly Hills, CA: Sage.

Media Matters (1986, July). Profiling the VCR market: A new study. p. 11.
Ogan, C. (1988). Media imperialism and the videocassette recorder: The case of Turkey. *Journal of Communication 38*(2), 93-106.
Peterson-Royce, A. (1982). *Ethnic identity.* Bloomington, IN: University of Indiana Press.
Roe, K. (1987). Adolescents' video use: A structural-cultural approach. *American Behavioral Scientist, 30*(5), 522-532.
Rose, P. (1981). *They and we: Racial and ethnic relations in the United States.* New York: Random House.
Snyder, S. (1985). Foreign video rentals popular. *Boston Globe,* 26 December.
Straubhaar, J., & Lin, C. (1986). *A quantitative analysis of the reason for VCR penetration worldwide.* Paper presented at the Telecommunication Policy Research Conference, Arlie, VA.
Subveri-Velez, F. (1986). The Mass Media and Ethnic Assimilation and Pluralism: A review and research proposal with special focus on Hispanics. *Communication Research 13*(1), 71-96.
Suraiya, B. (1983). India's dream merchants face up to a nightmare. *Far Eastern Economic Review, 122,* 80-81.
VCR usage on fast forward. (1988, April). *Broadcasting, 114*(14).
Video Marketing Newsletter 6(9). (1985, October).
Video News. (1986, 16 February). *Boston Globe.*
Vincent, J. (1974). The structuring of ethnicity. *Human Organization, 33*(4), 375-379.
Williams, F. (1987). *Technology and communication behavior.* Belmont, CA: Wadsworth.
Williams, F., Phillips, A., & Lum, P., (9185). Gratifications associated with new communications technologies. In K. E. Rosengren, L. Wenner, & P. Palmgreen (Eds.), *Media gratifications research* (pp. 241-252). Beverly Hills, CA: Sage.
Wilson, C., & Gutierrez, F. (1985). *Minorities and media: Diversity and the end of mass communication.* Newbury Park, CA: Sage.

12

VCR Narrowcasting in the Kibbutz

DOV SHINAR

Technology has played an increasingly significant role in the dialectics of developing and questioning communication philosophies, structures, and functions. Viewed in this perspective, the race for technological innovation in television, particularly in the last two decades, has been the focal point of the struggle for self-preservation on the part of profitable corporations and the pressures for devising alternative models.[1] Large screens, high definition images, improved color, and alphanumeric methods illustrate efforts to improve, rather than to change the product and service of big, centralized systems (Hurwitz,1978). VCR, videocassette, low-powered broadcasting, and other "small" technologies have been associated with change in communication models. Pre-meditated or coincidental, their development has triggered expectations and initiatives to use narrowcasting as a means to reinvent television (Kalba, 1979; Willener, 1976).

In philosophical and ideological terms, small media technologies promised to counterweight the inadequacies of the established models. Liberals and Marxists alike have admitted that wherever it has been used, television has divided communities between givers and receivers, programmers and watchers, distributors and consumers (Walsh,1980).

Some have seen these qualities as inherent to the medium (Groombridge, 1972; Mander, 1978), while others have been battling for the democratization and the demystification of television. In the framework of more comprehensive demands to democratize democracy, claims were made in the liberal camp for a broadcasting reform, intended to achieve a more equitable distribution of communication resources, curb manipulative control, and encourage citizens' participation and initiative (Henderson, 1974; Rogers, 1976; Branscomb & Savage, 1978; White, 1984). Others demanded to radically change the vertical, centralized, alienating nature of capitalist-controlled mediastructures into an emancipatory model, designed to enhance genuine relevance, popular involvement, and affirmative action (Enzensberger, 1972; Freire, 1974; Pasteka, 1979; Beltran, 1980). Accountability and public involvement were advocated as ways to demystify television. Their common philosophy has been "the right to communicate," which based on Article 19 of the Universal Declaration of Human Rights states that "everyone has the right of opinion and expression," including *association, information, and development* (MacBride, 1980: 173).

The concepts of *access, participation, and self-management* gradually emerged as practical expressions of the right to communicate. Access refers to the opportunities available to the public to choose programs, and to have a means of effective feedback to transmit its reactions and demands to communication organizations. Participation implies a higher level of involvement in the production process and in management and planning of communication services. Self-management is the most advanced form of participation, in which the public exercises the power of decision making, and is fully involved in the formulation of policies and plans (Jouet, 1977).

In the areas of policy and practice, this conceptual development resulted in a variety of alternative communication movements and experiments, in which advocates of radical change, proponents of moderate social decentralization, and social institutions—such as trade unions and religious organizations—were united in their opposition to the hegemony of the established media and in their support of various combinations of access, participation, and self-management. Small, portable, easy-to-operate inexpensive technologies seemed to be the perfect means for implementing their objectives.

In media research too, established models have been challenged by newly proposed relationships between product and process, source and recipient. Concepts such as channel combinations, functional interchangeability, and division-of-labor among different media emerged from findings on the limitations of individual media in satisfying communication needs (Katz & Gurevitch, 1976). Effect-oriented, product-based research (which accepted media premises that the program is the ultimate goal, and viewed planning/production/distribution processes as geared to their attainment) had to face the proposition implied in the right to communicate—that *participation* in all communication processes may be as important as the final product. Likewise, an alternative pattern was proposed to the traditional tight separation between production/transmission and consumption roles, in terms of less rigid boundaries between individual and community as recipient and as source (Willener, 1976).

Research problems typical of the established mass media—referring to control, structure, and resources; organization; operation and management; content and penetration; ethics and privacy; transfer/adoption of models; and legislation—appeared in different fashion, posed new theoretical and methodological challenges, and caused some researchers to ask whether the hitherto almost exclusive concern with the mass media has not become obsolete.

VCR AND BROADCASTING REFORM: REVIVED PROMISE AND DISENCHANTING PERFORMANCE

The advent of the VCR in the 1970s originally implied the possibility of genuine change. The flexibility of the new technology in terms of playback and off-air recording increased freedom in the choice of contents, distribution, and timing of use and reduced dependence on monopolistic broadcasting sources. In contrast with earlier versions of access, promoted earnestly or as lip-service by the established media, which did not imply relinquishing control (such as phone-in and audience-participation shows), the VCR triggered a true decentralization of control over television consumption. Advocates of television reform wanted more, however. Their search for grassroots models was geared to all television processes—planning, production, distribution, and consumption. They wanted to use video to enable individuals and groups to exercise greater control over the environment, to develop critical awareness and political organization, to create channels for local expression and interaction, and to foster cultural identity and artistic relevance.

For some ten years before the introduction of the VCR, activists had been experimenting with earlier generations of audio-visual technology—8mm and 16mm film and open-reel, low-gauge video—in order to attain these objectives. Following the Canadian "Challenge for Change" prototype, dozens of projects mushroomed around the globe. These were well-intentioned, idealistic, but sometimes naive endeavors to establish nonprofessional *community* television.[2] The fact that none of them succeeded has been attributed, in part, to the cumbersome and unreliable pre-VCR equipment (Worth & Adair, 1972; O'Sullivan-Ryan & Kaplun, 1980; Nigg & Wade, 1980; Walsh, 1980). Thus, it did not take long for grassroots communicators to perceive the promising prospects of the VCR technology. VCRs were easier to move and operate, cheaper, compatible with ENG and cable, offering a better storage capability than open-reel tapes, and easier to edit. A new technological promise thus served to revive the ideology of community television.

Unfortunately, the VCR experience has been as unrewarding as that of previous technologies. Few experiments to stimulate dormant civic feelings through community television have survived their euphoric infancy to become full fledged communication systems. True, some evaluations pointed to successful cases in North America and Europe (Berrigan, 1977, 1979; Brownstein, 1978; Burns & Elton, 1978; Moss, 1978; Lewis, 1984). These examples are, however, a marked minority among hundreds of experiments conducted in all parts of the world. In addition, success has usually referred to continued activity and growth, whereas "little research which could substantiate the claims of video in the area of social animation has been carried out . . ." (Berrigan, 1977: 211), leaving open the question of the net value of access and participation.

EXPLAINING THE FAILURE

The very promise of community television has thus remained ambiguous. As the improved technology did not provide satisfactory blueprints for a broadcast reform, new ideas were sought to replace the notion that appropriate technology would suffice to trigger a communications revolution. The resulting conceptual effort produced three variables, based on sociocultural rather than technological factors: the *compatibility* of the medium with the social context in which it is being applied; the *demand* for participatory communications; and the availability of adequate *resources* (Berrigan, 1977, 1979; Postgate, Lewis, & Southwood, 1979; Lewis, 1984).

The first variable (considered by diffusion researchers to be essential for effective adoption (Katz, 1973)), is the level of compatibility between an innovation and an adopting unit. Many experiments to establish small-scale participatory television revealed unsatisfactory compatibility levels: some researchers' claimed that media participation cannot be achieved without the parallel institution of participation in other spheres (Gonzaga-Motta, 1984), while others argued that community television was introduced in the industrialized world in neighborhoods whose population shared geographic proximity, but which did not satisfy basic sociological requirements of community (Mattelart & Piemme, 1980). The transplantation of models has triggered another compatibility problem: notwithstanding differences in structural, cultural, and technological contexts, in political tensions and in financial capabilities, many community television ventures in Europe, Oceania, and the Third World borrowed their conceptual premises from the United States and Canada, resulting in faulty adaptation (Berrigan, 1977; O'Sullivan-Ryan & Kaplun, 1980). The transplantation of the once fashionable Canadian "Challenge for Change" model to the Australian "Video-Access" project, for example, revealed incompatibilities between the imported emphasis on "process"—community members' involvement in planning, production, distribution, and utilization—and the final product, the program. Australian analysts found the Canadian emphasis on process incompatible with the Australian reality: "Process," they reported, "is only the philosophy of access providers. Access users must feel they are capable of making a 'good' product or their interest wanes" (Lewis, 1984: 76).

In the Third World, the open-society and equal rights premises underlying the imported models often proved incompatible with typical restrictions on decentralized public opinion mechanisms, with the low valuation placed on citizens, and with a frequently poor self-image (O'Sullivan-Ryan & Kaplun, 1980; Lewis, 1984).

The second variable—need and demand—highlights additional difficulties. A sociocultural resistance to participation was observed in most urban and rural experiments in North America and Latin America, as well as in Western Europe (Berrigan, 1977; Lewis, 1984; Atwood & Mattos, 1984), suggesting that either there was no need for community television at all, or that the application of the medium could not overcome local barriers, or worse, that the experimenters were not aware of the problem. Thus, the assumptions on communities' readiness to accept participatory television, and on their automatic adoption of simultaneous source/recipient roles revealed wishful thinking rather than proven fact.

Finally, the promise of community television has been found to be intimately related with the financial, technological, and human resources available for the development of both process and product. The lack of funds has been a universal constraint. In the Third World, community media have seldom been included in major financial priorities. Elsewhere, the costs of decentralization have always been a deterring factor, especially where they involved competition with commercial enterprise. Thus, experiments in Asian, African, and Latin American countries went through financial ordeals, along with Western European efforts, with Australian and even Japanese ventures and with many cases in the United States and Canada (Berrigan, 1977, 1979; Nigg & Wade, 1980; Fauconnier, 1984; Lewis, 1984; Prehn, 1985). Technology has been another source of constraint. Indeed, the advent of the VCR, computerized editing systems, and time-base correctors has eased production difficulties considerably, but these items have remained relatively expensive, have demanded moderately high skill levels, and have not been available to most community groups. Manpower and training constraints have included everywhere difficulties in recruiting and training community groups in technical, artistic, journalistic, and human relations skills; in developing adequate leadership; and in mobilizing back-up support (Berrigan, 1977; Nigg & Wade, 1980; Lewis, 1984).

VCR NARROWCASTING IN THE KIBBUTZ: THE PROMISE

The introduction in the late 1970s of narrowcast television in the Israeli *kibbutz* (collective rural community) provided an opportunity to clarify the conditions necessary for achieving the objectives of community television. Unlike the small-scale, on-again off-again nature of most previous experiments, the sizeable acquisition and application of narrowcasting equipment, and the expected presence of compatibility, demand, and resources promised to satisfy substantive and methodological requirements. Thus, out of a sample of 71 kibbutzim (the plural of kibbutz in Hebrew) with cable facilities, five systems were introduced in 1978-1980, 45 in 1981-1984, and 21 in 1985 (Shinar & Shur, 1986).

This chapter reports a study undertaken between 1985 and 1986, under the auspices of the Yad Tabenkin Institute for Kibbutz Studies, aimed at exploring the promise and performance of kibbutz cable television as a community medium and at making policy recommendations. At the same time the study—including a "Kibbutz Video Survey," content analysis of programs, and three in-depth case-studies—was considered useful to learn about the more general conditions necessary for the viability and effectiveness of community television.

The Kibbutz Video Survey (Shinar & Shur, 1986) included a detailed questionnaire on the introduction, operation, and resources of local cable stations, sent to informants in all 263 kibbutzim in Israel. Completed questionnaires were returned by 182 informants, featuring a fair representation of the total distribution of kibbutzim by size, age and ideological orientation. Content analysis was used to investigate 33 local programs produced in nine kibbutzim and all items of the first seven editions of "Arutz Lakibbutz" (Kibbutz Channel), an overall kibbutz movement newscast screened in 98% of the kibbutzim surveyed, through local cable systems or direct closed-circuit. The in-depth case studies included interviews with kibbutz members and with key figures in kibbutz television, as well as an analysis of relevant documents. In 1985 at least 121 (40%) of the 263 kibbutzim owned one VCR or more. At least 52 kibbutzim owned cameras and 22 owned electronic editing equipment. VCRs were being used for overall community purposes in 71 kibbutzim, while local cable stations enabled members of at least 52 communities to view rented and self-produced programs in their homes. Direct closed-circuit screenings were being conducted in an additional 14 communities, with local schools and special interest groups using video in another 50 kibbutzim. In an addition, two fully equipped and staffed central units were operated by the nationwide kibbutz organizations for the production and distribution of programs to their affiliates, and four regional centers were active in training and production (Shinar & Shur, 1986).

Compared to other cases, kibbutzim were found to satisfy most requirements identified for community television. Unlike many other earlier sites for community television projects, kibbutzim are full-fledged communities, not only in terms of geographic proximity, but also in shared values, symbols, and life-styles. Kibbutz life is based on voluntary selectively accepted membership, pragmatically engineered according to a combination of socialist and Zionist ideologies. Notwithstanding differences between individual communities (Krausz, 1983), the kibbutz has been found to provide the motivation, participation, and mutual responsibility necessary for the formation and persistence of community. Production and consumption, for example, are collective. Most members work together in agriculture, industry, and services. All members receive goods and services from the kibbutz via collective institutions such a communal dining hall, laundry, and clothing store. The socioeconomic status of kibbutz members is not directly related to the goods and services they receive or produce. Unlike other systems, the farm manager, the head of the industrial plant, or the highest officials receive no more than other members; status is achieved not through seniority, wealth, or family rights, but through the contribution to the community. Finally, in addition to internal kibbutz relations, the concept of community also applies to interkibbutz dimensions, whereby the affiliation of individual kibbutzim to the central organizations that form the Kibbutz Movement Federation provides ideological, cultural, and practical cooperation.

The introduction of community television in the kibbutz displayed a significant awareness of model-transplantation problems. The point of departure for recommending the adoption of the medium was a search for the solution of well-defined kibbutz problems. The first policy paper issued in 1981 by the Kibbutz Movement's Communications Committee suggested that the very existence of the kibbutz as such depends on the active involvement of its members in all spheres of life. The efforts to maintain and promote active involvement were guided by the premise that as a unique way of life, the kibbutz has been compelled since its early days to look for compatible answers for its problems, through local experimentation rather than through the adoption of models developed elsewhere. Thus, community television was considered a possible means to promote involvement by stimulating public dialogue and reaction to local, regional, and national issues, by encouraging members to express their views and concerns, and by bringing up matters to public knowledge and public discussion (Kibbutz Communications Committee, 1981).

The position taken by kibbutz media activists in the "process versus product" and "providers versus users" dilemmas was that while the model being locally devised could not function without a strong process component, attention should be given to producing a viewable and meaningful product, together with an awareness of the fallacy involved in trying to imitate broadcast television (Tamir, 1985).

The second variable identified in community television research—demand—is present in the kibbutz reality almost by definition. The equality of rights and duties, the mutual dependence among members, and the principles of direct democracy are but a few sources of this need for participation and communication in kibbutz life. They are complemented by the fact that kibbutzim are not isolated communities. Although making up no more than 3% of Israel's population, active involvement in national affairs has been a major tradition of the movement, as expressed by the role of kibbutz members in the struggle for the establishment of the state and of the country's public life, security affairs, economy, and culture. A steady flow of communications is necessary for personal and institutional functioning as well as for maintaining morale, motivation, involvement and solidarity. In addition, the demand for information has been strongly emphasized by the economic and occupational changes which have affected the kibbutz in recent years. Technological development, professional advancement, and economic growth have increased the need for internal and external communications. This has been a complex task for kibbutz members and institutions, with implications for the comprehension of a changing environment, accessibility to events, and exertion of the right to participate in decision making (Katz & Golomb, 1983).

Changes in the collective ideology have been an additional source of demand for participatory communications. Kibbutzim have developed from isolated groups of young, pioneering idealists to communities of several hundreds of members and a wide range of personalities, interests, and skills. During almost one century of existence, natural erosion affected ideological zeal and collective orientations, coupled with the impact of a rising standard of living and a diminishing isolation from the urban middle class. Personal property has been increasingly tolerated, revealing a more liberal attitude to the formerly rigid concept of equality. The family unit has been competing with the hitherto predominant community orientation, as illustrated by the decision adopted by most kibbutzim to have the children spend the night at home instead of at the communal house and by families taking some of their meals at home instead of at the communal dining hall. These trends of affluence have exposed the kibbutz collective ideology to threatening influences. One generally approved conclusion called for strengthening community ties through the activation of communication channels tailored to local needs (Magen, 1985).

Field data from the Kibbutz Video Survey support this conclusion: about 2/3 of the informants in 171 kibbutzim (63%) reported the awareness of the need for improved communications in their communities—40% among a majority of members and 23% among part of the membership. Economic and financial information were the leading topics in demand. A greater demand was found for internal rather than external communications. Most communities reported being well-provided by the mass media with information on the world, the region, and the country, while the demand for communications with non-kibbutz communities was found to be lower than the need for communications on kibbutz cultural continuity, local artistic expression, and links between kibbutz members and the movement. The majority of respondents proposed to introduce local television (79.6%), to improve the local newsletter (81.7%), and to intensify interpersonal relations (79.2%) as ways to meet the demand.

Kibbutz television has been fortunate in the third potential problem-area: resources. In the late 1970s, when cable television was first introduced in the movement, there were no real problems of technology, funding, and manpower. At that time major technological difficulties had already been solved and became a problem of funding and skill. Also during those years, kibbutzim were experiencing an unprecedented financial growth. No difficulties were found in installing transmission cable systems (usually a by-product of the installation of telephone lines) nor in purchasing production and playback equipment, sometimes of a quality far exceeding the needs of community television. During the first five years of kibbutz television, the overall sum invested by kibbutzim in the purchase of VCRs and other production/playback equipment was estimated at three million U.S. dollars. This sum is about three times higher than the budget allocated by all kibbutzim to their members for the purchase of books.[3] In the area of manpower, the kibbutz has earned a reputation for problem-solving inventiveness and for the ability to perform cooperative assignments. Thus, the demands posed by community television for positive attitudes to innovation and participation as well as for autonomous initiative and decentralized structure were met by a favorable predisposition.

Compared with findings on the destructive influence of external initiative, excessively centralized authority, and governmental control (Berrigan, 1977; Lewis, 1984), the promise of community television in the kibbutz was brighter than elsewhere.

Compared with the controversial introduction of television sets in the kibbutz movement in the late 1960s, the diffusion of VCR narrowcasting equipment in the late 1970s indicates a stronger acceptance of video, notwithstanding obvious differences in pace and proportion.[4] Two different patterns emerge from the comparison of data gathered for the diffusion of television sets (Gurevitch & Loevy, 1972) and of VCR equipment (Shinar & Shur, 1986) in the kibbutz, in terms of the percentages of kibbutzim where both types of items were available at the time of each study (third measuring), twelve months earlier (second measuring), and at the time of their introduction (first measuring). The diffusion pattern of television sets displays an initially modest rate of growth (5.6%), from 19.5% in the first measuring to 25.1% in the

Table 12.1 Diffusion of Television and Community VCR in the Kibbutz Movement (percentages of kibbutzim)

	Television *(N = 231 kibbutzim)*	*VCR* *(N = 263 kibbutzim)*
First measuring	(April 1968)	(mid-1980)
Early introduction	19.5%	1.9%
	(45)	(5)
Second measuring	(June 1968)	(mid-1984)
12 months preceding study	25.1%	19.0%
	(58)	(50)
Third measuring	(June 1969)	(mid-1985)
At time of study	61.0%	27.0%
	(141)	(71)
Percentage of growth between first and second measurings	5.6%	17.1%
Percentage of growth between second and third measurings	35.9%	8.0%

second. This was followed by a dramatic increase (35.9%) to 61% twelve months later, suggesting that an uphill path led to opening the way for large-scale adoption. Indeed, the introduction of television in the kibbutz was met by strong resistance, on the grounds of its alleged destructive influence on the collective, participatory, and local creativity ethos (Gurevitch & Loevy, 1972). The diffusion of VCR and other narrowcasting equipment indicates, in contrast, an initially high rate of adoption (17.1%), from 1.9% in the first measuring to 19% in the second. A modest increase (8%) to 27% took place in the following twelve months, suggesting that the VCR had to overcome fewer early obstacles, but that some difficulties did emerge to slow the pace of diffusion.

THE PERFORMANCE: PROCESS

Trial and error experiments, coupled with uncoordinated efforts were typical of the performance of kibbutz television in its first formative years. Sixty-nine percent of the local stations (47 out of 69 cases) were introduced as a result of individual or group initiative rather than of collective resolutions taken by kibbutz communal institutions. Public discussion on the introduction of cable television took place in the general assemblies of no more than 23 kibbutzim, one-third of the cases studied. The guidelines formulated by kibbutz communication activists did not reach the local level or were rejected altogether.

About half of the 241 television crew members for whom data were available in the Kibbutz Video Survey, had no training at all; about one quarter were given only basic training in kibbutz movement crash courses, while another quarter had longer training. Local operations were run by staffs with an average size of five members, with a range from three to eight. About half of the staffs (23 out of 53 for whom data were available) had been working without any systematic procedures, 19 crews formulated general guidelines, while only seven crews had overall policies. Preproduction meetings were never conducted in more than half of the kibbutzim and only 20% had regular staff meetings. Decisions on production and distribution are made by the station staff alone in 21 sites, by the staff jointly with kibbutz committee members in 22, and by the staff jointly with a specially elected committee in 10 kibbutzim. Work with video productions usually is conducted at the expense of staff members' free time. Only 20% of the crews said they had adequate time for planning, production, and distribution assignments. Likewise, most stations lacked autonomous budgets and depended on the kibbutz' cultural committee or secretariat for funds. Nevertheless, few complaints were reported on the application of censorship through this dependence. In fact, almost half of the stations reported on their satisfaction with current budgetary arrangements. The major basis of control has been public opinion, usually through informal criticism.

Planning and coordination of production and transmission have been as casual as station operation. The objectives for the introduction of community television, formulated in 69 kibbutzim with local stations, do not display a high degree of compatibility with the areas rated in 171 kibbutzim as highest in perceived need of communication improvements (see Table 12.2). Economic and financial information were rated as priority areas for the improvement of local communications, but were assigned much lower priority among kibbutz television objectives. Entertainment was not mentioned at all as a priority area for communications improvement, but more than 40% of the local stations have had it as a major programming item. Education and culture were given low priorities among the areas of need, but were considered the most important objectives for the introduction of television.

Table 12.2 Kibbutz Communications Demand and Kibbutz Television Supply

Demand				Supply			
Areas in which the need for communication improvements has been most acutely felt (% kibbutzim) N = 171		Objective for introduction of cable television (% kibbutzim) N = 69		Major emphases in kibbutz television local transmissions (% kibbutzim) N = 69		Major topics in kibbutz television local productions (% kibbutzim) N = 175	Major topics in "Arutz Lakitbutz" (Kibbutz Channel) (% kibbutzim) N = 51
Economics	73%	Social, educational, cultural improvement	72%	Entertainment	42%	Current events 32%	Culture, society, education, heritage 45%
Finance	70%	Documentation	60%	Involvement	30%	Economic reports 30%	Kibbutz external relations 29%
Society	62%	Communications improvement	57%	Information	14%	Daily problems 18%	Economics, finance 18%
Education	55%	Economic improvement	57%	Documentation	14%	Trivia 15%	Welfare 8%
Culture	55%	Democratization	55%			Welfare 5%	
Welfare	48%	Entertainment	75%				

In addition, the strong community orientation of kibbutz video activists in the initial days does not appear to have reached either the field or the leadership. The failure of kibbutz television founders to persuade the movement's central bodies even to discuss the official adoption and recommendation of the community television model obviously reduced the model's legitimacy and acceptance at the local level.[5] Local crews have been only sparsely guided by central units. The available guidance, given mainly in crash courses, included orientation on the purchase and operation of equipment and production work, with little concern about the basics of journalism, station development, and community activation. Ideological, operational, or professional guidelines have not been provided, given that blueprints on these matters were never developed at the central level. This is hardly surprising, since production has become the major objective at a relatively early stage. Central units became staffed with trained crews, their programs have been recorded on 3/4 inch high-band tape (in order to achieve broadcast quality, which could enable them to market their product to professional television), and assistance of outside contractors has been sought in editing the final product. Following broadcast television standard practices, productions of local crews have seldom been considered qualified to be included in the "Arutz Lakibbutz" (Kibbutz Channel) movement-wide periodical news magazine, and decision making has been highly centralized.

THE PRODUCT

Although there is some locally produced video, loaned, rented, and purchased materials make up the major part of kibbutz TV programming, a pattern not unlike the use of home-VCR. Programs produced by the kibbutz movement's central units—particularly Arutz Lakibbutz—have been transmitted in all 69 community stations studied. In addition, programming in 59 stations has included off-air recordings from Israel Television, mainly to compensate for the overlap in scheduling between kibbutz communal functions and popular television programs. The most outstanding example has been occurring on Saturday nights, when kibbutz general assemblies coincide with the broadcast of the major weekly sports program. Rented feature films have been another popular item in 54 stations, followed in 32 kibbutzim by off-air recordings from Israel Instructional Television.

Table 12.3 Programs Transmitted in Kibbutz Television [n = 69 kibbutzim]

	% Kibbutzim
Import from kibbutz movement: Arutz Lakibbutz (kibbutz channel)	98.5
Recordings from Israel TV	85.5
Rented feature films	78.2
Recordings from Israel instructional television	46.3
Locally produced features	46.3
Locally produced newscasts	26.0
Preparation for coming general assemblies	18.8
Games and specials	14.4
Electronic bulletin board	8.6

Local productions, recorded, edited, and transmitted on 1/2 inch VCR (except for 9 stations with 3/4 inch U-matic equipment) were less widespread. Features on local topics have been transmitted on irregular schedules in 32 kibbutzim and periodical local news magazines in 18. Less common local programs include items to be discussed in coming general assemblies, electronic bulletin boards, games and contests, productions for kibbutz festivities, and live coverage of kibbutz committee meetings.

Kibbutz productions, central and local alike, have made up only a small part of the transmission schedules. With an average duration of 20 to 30 minutes, local programs have been transmitted once in three to four weeks, and kibbutz movement programs once in four to six weeks. Local stations have emphasized current events and information, with particular reference to social issues. A typical local news magazine may include a discussion on a topical issue such as the composition of nursery school age-groups; whether or not to conduct BarMizvah celebrations; items from the field, such as dog-life in the kibbutz, fire-prevention, etc., items on kibbutz local artists; items on the local industrial plant, agricultural farm, etc.; financial reports; and variety items, such as celebrations, and humorous events. Special programs have dealt with recognized problems, such as the adherence to cults, kibbutz members returning from army service, and documentation of festive, political, and other events. Arutz Lakibbutz has dealt with kibbutz identity (such as the role of the kibbutz in Israeli society, moral/ideological stocktaking); internal mechanics (such as innovations in education, the introduction of an automatic banking machine, and budgetary cuts); current information, and arts/entertainment topics. Along with this service, the public relations function of the program has been clearly shown in the preference for safe topics, such as cultural and social reports (45% of the items) rather than economic and financial controversy (18% of the items, notwithstanding a much higher audience demand); and in the preference for covering kibbutz official institutions (46% of the items) rather than events (26%) and individuals (19%).

Productions of kibbutz movement central units have generally received greater human, financial, organizational, and technical resources than local productions. However, their tendency to imitate broadcast television in contents and formats, and the attempts to assume critical positions toward the kibbutz establishment have been incompatible with their didactic, mobilized tone. Formality and pompousness were identified in more than 80% of the items analyzed. Messages have been usually positive, expressing praise, pride, and confidence in more than 50% of the items; consensus and some criticism in 45% to 50% of the items; and negative feelings, including frustration, doubt, and mistrust in 10% to 30% of the messages transmitted. A typical Arutz Lakibbutz program may include: items on the contribution of the movement to Israeli society, items on current information (such as the agricultural season, the situation of kibbutz industrial plants), items of general interest (i.e., educational innovations, construction, movement's staff returning home), and entertainment items.

The influence of broadcast television has been apparent in Arutz Lakibbutz and in local productions alike. In the former it has been expressed in the visual treatment (sets, background, performing, and announcing style) and in verbal and symbolic language, including artwork. In the latter this influence can be recognized in the imitation of program structures and formats. Local programs, however, have been typified also by some conscious or accidental attempts to develop a specific style, illustrated by the lack of manipulation of reality and by the development of a rhetoric of intimacy, expressed in an informal familiar tone, and in the fact that characters do not have to be identified for the viewer.

EFFICIENCY AND EFFECTIVENESS

The performance of kibbutz television as a community medium has not yet measured up to its bright promise. In terms of efficiency, most local stations have displayed a casual approach to planning, production, distribution, and feedback. The central units have become increasingly centralized and alienated from the field. A tendency toward professional broadcasting rather than to community narrowcasting is apparent in both cases.

In terms of effectiveness in goal achievement, access has been selective, having provided little opportunity to kibbutz members to come closer to the system. Although kibbutz members were consulted on transmission schedules, they have had little influence over the selection of imported and locally produced programs. Access has been even more limited at the central unit level, expressed by a minimal share of local crews (and none on the part of individual members) in the selection of materials and feedback.

Participation has been as limited as access, with little involvement of the public in planning processes and decision making. Such participation has included, at best, voting in the general assembly for or against the allocation of funds to the local crew. There has been no public participation in central units' decision making. Moreover, few conscious efforts were reported on the recruitment and encouragement of large numbers of kibbutz members to participate in program production. Out of 48 kibbutzim for which data were available on this matter in the Kibbutz Video Survey, 26 reported strong interest and 17 reported some interest by members in being involved in the work of the local station. Explicit interest in participating in production was expressed by 28 of the 66 respondents in one of the in-depth case studies; by 24 out of 110 respondents in the second case study; and by 28 out of 200 respondents in the third case. This amount of demand is certainly not matched by the 5-member average size of crews in the 69 stations studied nor by the average number of participants in the production of the 33 local programs analyzed (3.5 participants per kibbutz in 9 communities with an average population of 200 adults each).

One consequence of these shortcomings has been the virtual absence of self-management. Kibbutz television has come to be considered as a hobby of a few video activists in local stations and as the privilege of professionals in the central units.

Conclusions: The Missing Variables

The faulty performance of kibbutz cable television cannot be ascribed to the absence of basic prerequisites, compatibility, demand,

and resources, since kibbutzim have been generously equipped with them. The findings of the present study point to other explanations. First, the formative phase of kibbutz television was not guided by a clear conceptual framework. The efforts invested since the late 1970's by the movement's community television founding fathers in devising applicable blueprints did not find significant expression in the field. The ineffective diffusion of community television principles among crews and audiences reflects the lack of policy and the weakness of dissemination channels. These have been powerful obstacles to the attitudinal and behavioral charge required by the adoption of participatory media. The adherence to the professional model—in which only one component of the community media model, namely production, is the major objective—was not free from the commitment to use television for sociocultural improvement. Thus neither access and participation nor centralized specialization and professionalism became full and legitimate objectives. The confusion between elements from both models has prevented achieving most of their goals.

Problems of systemic balance provide a second explanation. A stronger emphasis has been given to certain components of television systems (such as transmission and production equipment and imported programming) than to other components, including planning, training in journalism practices and community activation, and developing operational blueprints.

Finally, the lack of recognition of community television as a specific medium provides an additional explanation. The view of local community stations as miniature broadcasting organizations, implied in the overemphasis given to production, has meant giving up the particular qualities of community television. In this sense, the difference between the professional and community models is that the former promotes essential norms (such as planning and creativity, correct camera work, and high-quality editing) as ultimate goals, while the latter prescribes the use of these norms for the achievement of social and cultural objectives, including the promotion of community self-expression and authenticity, the opening of channels for public involvement, and multidirectional communication avenues.

Since the completion of the study in 1986, no significant changes have been reported in the situation depicted by the study. A recession in kibbutz economy has caused the discontinuation of subsidized productions. Arutz Lakibbutz has further emphasized its professional aspirations, mainly through offering advertising services to kibbutz-based industrial plants, in order to compensate for the loss of income.

At the local level, less than a dozen new stations became operative. The movement's leadership has not acknowledged the participatory model nor has it encouraged its development.

Thus, notwithstanding the presence of most prerequisites identified in previous research, kibbutz television has displayed acute performance problems, notably a low level of institutionalization, consistent lack of planning, and a confusion of models. Consequently, compatibility, demand, and resources can be considered *necessary* conditions, without which community television is doomed from the outset. The variables emerging from the analysis of kibbutz television—conceptual framework, systematic balance, and recognition of the medium's specific qualities—may be considered at least as some of the *sufficient* conditions for the effective application of community television.

NOTES

1. Research data indicate, for example, that Third World governments that embarked in "big technology" investments abandoned regional provision in broadcasting and the press, and that almost everywhere local and regional media operators have been pressured to upgrade their equipment in order to compete with big media organizations (Lewis, 1984: 7).

2. "Community television" is used here to mean "a station characterized by the local nature of its ownership, programming, and the market it is designed to serve . . . that provides for management, operation, and programming primarily by and for the members of the local community . . . (and that) should allow for community access and reflect the interests and special needs of the audience . . . ." See Canadian Radio-Television and Telecommunications Commission (1986). Public Notice CRTC 1986-176 proposed Regulations Respecting Television Broadcasting, pp. 18-19.

3. The average annual budget for the purchase of books allocated to each kibbutz member is 10 shekels (Israel) or U.S. $6.60. Multiplied by 30,000 members, it amounts to some U.S. $200,000 annually and U.S. $1 million in five years. In addition to this earmarked allocation, members use funds from their general personal budgets for purchasing books.

4. Differences in absolute pace and magnitude of adoption and diffusion can be accounted for by the easier installation and use of television sets, and the more complex nature of narrowcasting systems (infrastructure, cost, training, and organization).

5. This unfavorable background for participatory media suffered a further setback when, following the rotation principle in the staffing of central positions in the movement, time came for the group of community television initiators to go back home. They were relieved in late 1982 by a team of more professionally inclined communicators, who concentrated on the promotion of centralized production. Although this team was relieved by another in 1986, the professional inclination is still predominant in 1988.

REFERENCES

Atwood, R., & Mattos, S. (1984). Mass media reform and social change: The Peruvian experience. In G. Gerbner and M. Siefert (Eds.), *World communication: A handbook*. New York: Longman.

Beltran, L. R. (1980). Farewell to Aristotle: Horizontal communication. *Communication* 5(1).

Berrigan, F. (Ed.). (1977). *Access: Some western models of community media*. Paris: Unesco.

Berrigan, F. (1979). *Community communication: The role of community media in development*. Paris: UNESCO.

Branscomb, A., & Savage, M. (1978). The broadcasting reform movement: At the crossroads. *Journal of Communication 28*(2).

Brownstein, C. N. (1978). Interactive TV and social services. *Journal of Communication 28*(2).

Burns, R., & Elton, L. (1978). Programming for the future. *Journal of Communication 28*(2).

Enzensberger, H. M. (1972). Constituents of a theory of the media. In D. McQuail (Ed.), *Sociology of mass communications*. Middlesex, England: Penguin.

Fauconnier, G. (1984). Serving two cultures: Local media in Belgium, In G. Gerbner & M. Siefert (Eds.), *World communication: A handbook*. New York: Longman.

Freire, P. (1974). *Pedagogy of the oppressed*. New York: Seabury.

Gonzaga-Motta, L. (1984). National communications policies: Grass roots alternatives, In G. Gerbner & M. Siefert (Eds.), *World Communication: A handbook*. New York: Longman.

Groombridge, B. (1972). *Television and the people*. London: Penguin.

Gurevitch, M., & Loevy, Z. (1972). The diffusion of television as an innovation: The case of the kibbutz. *Human Relations 25*(3).

Henderson, H. (1974, March). Information and the new movements for citizenship participation. *Annals of the American Academy of Political and Social Sciences*.

Hurwitz, S. (1978). The corporate role. *Journal of Communication, 28* (2).

Jouet, A. (1977). *Community media and development*. Working paper for Belgrade meeting. Paris: UNESCO.

Kalba, C. (1979). The video implosion: Models for re-inventing television. In R. Adler & W. Baer (Eds.), *The electronic box office*. New York: Praeger.

Katz, D., & Golomb, N. (1983). Integration, effectiveness and adaptation in social systems. In E. Krausz (Ed.), *The sociology of the kibbutz*. New Brunswick, NJ: Transaction.

Katz, E. (1973). *On the use of the concept of compatibility in research on the diffusion of innovation*. Jerusalem: The Israel Academy of Sciences and Humanities Proceedings V(5).

Katz, E., & Gurevitch, M. (1976). *The secularization of leisure*. London: Faber and Faber.

Kibbutz Communications Committee (1981). *Kibbutz communications in the age of electronic media*. Working paper. Tel Aviv: Kibbutz Movement Federation. (Hebrew).

Krausz, E. (Ed.). (1983). *The sociology of the kibbutz*. New Brunswick, NJ: Transaction.

Lewis, P. M. (1984). *Media for people in the cities: A study of community media in the urban context*. Paris: UNESCO.

MacBride, S. (1980). *Many voices, one world*. Paris: UNESCO.

Magen, A. (Ed.). (1985). *Communications technology in the kibbutz*. Tel Aviv, Israel: Yad Tabenkin Institute for Kibbutz Studies. (Hebrew).

Mander, J. (1978). *Four arguments for the elimination of television*. New York: Morrow.

Mattelart A., & Piemme, J. M. (1980). New means of communication: New questions for the left. *Media Culture and Society, 2*.

Moss, M. L. (1978). Research on community uses. *Journal of Communication, 28*(2).

Nigg, H., & Wade, J. (1980). *Community media*. Zurich, Switzerland: Regenbogen-Verlag.

O'Sullivan-Ryan, J., & Kaplun, M. (1980). *Communication methods to promote grassroots participation*. Paris: UNESCO.

Pasteka, J. (1979). *Right to communicate: A socialist approach*. Paris: Unesco.

Postgate, R., Lewis, P. M., & Southwood, W. A. (1979). *Low cost communication systems for educational and development purposes in third world countries*. Paris: UNESCO.

Prehn, O. (1985). *Community television in Denmark*. Aalborg: University of Aalborg.

Rogers, E. (1976). Communications and development: The passing of the dominant paradigm. *Communication Research 3*(2).

Shinar, D., & Shur, S. (1986). *Kibbutz video survey*. Tel Aviv, Israel: Kibbutz Movement Federation. (Hebrew).

Tamir, A. (1985). *Closed circuit: Video in the service of the kibbutz*. Tel Aviv, Israel: Sifriat Poalim. (Hebrew).

Walsh, B. (1979-80, December-January). Community television. *Cinema Papers*.

White, R. A. (1984). Communication strategies for social change: National TV versus local public radio. In G. Gerbner & M. Siefert (Eds.), *World Communications: A Handbook*. New York: Longman.

Willener, A. (1976). *Videology and utopia: Explorations in a new medium*. London: Routledge and Kegan Paul.

Worth, S., & Adair, J. (1972). *Through Navajo eyes: An exploration in film communication and anthropology*. Bloomington: Indiana University Press.

13

The Worldwide Cultural and Economic Impact of Video[1]

CHRISTINE OGAN

The videocassette recorder is not just another Japanese electronic media toy to be used by affluent Americans. In fact, Americans have been slow to buy and use the VCR compared to the rest of the world. It has had a major impact on other cultures, especially those where the mass media entertainment choices are few and/or where the information sources are controlled. According to one source, the VCR was expected to have spread to at least half of the households in at least 21 countries, including 12 developing countries by the end of 1988 ("Mid-year video review," 1988). The VCR is used for time shifting and the viewing of domestic and foreign films, but also for viewing family celebrations and politically forbidden information. It has had an international impact far beyond what forecasters expected from a technology that could only transmit content—and content that was primarily designed for transmission by more powerful mass media. The VCR may not have been taken seriously because it has been considered an individual medium rather than a mass medium.

The videocassette recorder is one of several new media delivery technologies to affect cultural industries worldwide. These technologies have provided the means for a greatly expanded number of channels of content, made it more convenient for viewers to select video content without having to leave their homes, and provided an opportunity for program providers to move into markets never before open to them. As a result, the television and film viewing experience in virtually every country of the world has forever changed.

It is important to look at that change on both cultural and economic levels. Since much of the content that circulates internationally for playback on VRCs originates in the United States or western Europe, concerns over widespread cultural imperialism have been expressed. Content that could once be centrally controlled by television station or cinema owners began to circulate freely in places such as the Soviet Union (see Boyd, 1988), Saudi Arabia and Pakistan. If the cultural impact was exaggerated (see Ogan, 1988b), the economic impact has been evident. The U.S. film majors made more profits from video cassettes than from films in 1986. At the same time money was lost by both the U.S. majors and the many film producers all over the world to international video pirates.

This chapter will assess the economic impact of the VCR on cultural industries in the United States, western Europe, and the Third World. Based on data gathered on 43 countries, hypotheses dealing with the causes of piracy of videotaped content and effective policies for its reduction will be tested. The chapter will also discuss the nature of the concerns over the preservation of cultures in the face of increased U.S.-produced content and the status of the research that assesses the reality of those concerns.

INTERNATIONAL ECONOMICS OF THE VCR

The VCR has been a popular technology, diffusing more rapidly in countries with low media diversity than in the United States, where multiple cable channels, subscription television services, and wide availability of local cinema have been the norm for several years (see Ogan, 1985b). The rapid VCR diffusion took place in poor countries of the Third World and in western Europe at the beginning of this decade when most Americans viewed the technology as one that cost more than it was worth. By the mid-1980s, when prices dropped in the United States, the VCR became the most popular holiday gift item and penetration passed the 50% mark by 1987.

Unlike the purchase of a television or cable service, the VCR requires the viewer to obtain software for its useful operation. In the United States, the first VCR owners found sufficient viewing content on available over-the-air and cable television channels, so they used their VCRs for time shifting, thereby maximizing viewing choices. (Levy, 1980, 1983). But in countries where VCR owners had few television options, consumers sought content from videocassette rental stores, where they found foreign and domestic films and television programs in abundance. Now, of course, rental shops are quite popular in the United States too, and according to a 1988 study, VCR owners spend more time viewing prerecorded tapes than home-recorded programs (AGB Research, 1988). Most of prerecorded content rented to viewers has been obtained legally by the store owners in the United States, but in numerous countries much of the content has been pirated from both foreign and domestic sources.

The economic base of the video business is built on sales of hardware, and sale and rental of software. This chapter will not directly address the hardware or VCR industry. This industry is almost entirely controlled by Japanese companies, although South Korean and a couple of European companies are advancing on the Japanese.[2] We can mark the growth in profits of this industry by examining worldwide penetration rates. The major difficulty with penetration rate data is its inaccuracy. Even in western Europe, where controls on sales and counting procedures are well developed, estimates on penetration vary greatly. (Ogan, 1985b; also, see Table 13.1). Recent Gallup poll data has put the U.S. penetration rate at 65%, about 9% higher than the Electronic Industries Association had estimated in June 1988. Two other estimates for April 1988 were 54.5% and 56.6% ("Gallup poll shows," 1988). And a November 1987 survey by the American Video Association estimated the U.S. penetration rate at 62% ("Survey backs video erosion," 1988).

In some Third World countries, where VCRs are often smuggled in to avoid excise taxes, estimates are more suspect. Add to this problem the diffusion of VCRs carrying illegal Japanese or other trademarks and the penetration assessment becomes even more difficult.

The more interesting, and possibly the more important, economic issues are raised by the diffusion of prerecorded content on cassettes. Whether legally or illegally distributed, this content is having a major economic impact on television program and film industries in every country. I will discuss this impact without regard to the legality of distribution first, and then deal with the pirated content as a separate issue.

Declining Cinema Box Office. Although it is difficult to separate the influence of cable- and satellite-delivered content from that of the VCR, it is true that most countries have experienced major declines in box office revenues since the new delivery technologies became prominent. That phenomenon has been observed in the West as well as in the

Table 13.1 Videocassette Recorder Penetration Figures From Three Sources (in thousands)

Country	TV World (May 1988)	Screen Digest (November 1987) (projected to 1/88)	MPAA (6/21/88)
Argentina	226	325	385
Australia	3,300	3,436	3,035
Austria	330	535	632
Bahrain	78	83	87
Belgium	950	724	759
Bermuda	22	33	36
Brazil	1,400	1,750	3,000
Bulgaria	40	60	73
Canada	3,700	4,800	8,495
Chile	111	150	175
China	200	570	680
Colombia	500	615	670
Cyprus	36.5	47.6	100
Denmark	550	654	700
Ecuador	62.5	74	80
Egypt	601	605	705
Finland	600	591	665
France	4,500	6,021	6,855
German Federal Republic	8,700	10,017	10,259
Greece	250	545	590
Hong Kong	450	944	825
Hungary	200	190	235
Iceland	50	40	47
India	1,300	1,506	1,610
Indonesia	735	800	890
Iran	455.5	515	540
Iraq	175	195	205
Ireland	270	415	430
Israel	564	434	434
Italy	1,800	2,023	2,493
Japan	9,900	25,520	25,520
Jordan	61	72	80
Kenya	21	26	31
Korea, South	—	1,850	1,980
Kuwait	500	467	467
Lebanon	400	421	421
Luxembourg	31.5	42.2	40
Malaysia	1,502	688	730
Mexico	900	1,215	1,400
Netherlands	2,100	2,034	2,135
Netherland Antilles	—	36	36
New Zealand	360	510	950
Nigeria	190	406	505

(continued)

Table 13.1. (continued)

Country	TV World (May 1988)	Screen Digest (November 1987) (projected to 1/88)	MPAA (6/21/88)
Norway	600	528	550
Oman	50	44	44
Panama	92	155	80
Peru	250	310	310
Philippines	506	566	1,000
Poland	500	590	595
Portugal	405	604	604
Qatar	85	89	89
Saudia Arabia	2,005	1,368	1,368
Singapore	350	214	210
South Africa	510	535	580
Spain	3,000	2,848	3,330
Sweden	1,250	1,096	1,185
Switzerland	680	673	750
Taiwan	—	1,626	1,790
Thailand	420	520	570
Turkey	2,500	1,575	2,500
Uruguay	25.5	20	28
United Arab Emirates	100	86	98
United Kingdom	12,800	11,531	10,500
United States	41,000	50,870	50,180
USSR	100	575	675
Venezuela	800	794	840
Zimbabwe	14.5	24	32

Third World. Though VCR use is more widespread than that of cable or satellite, the influence of video cannot be independently determined. It would be even more difficult to establish a causal link between VCR use and declining box office revenues. No country-by-country data on box ofice revenues are readily available for analysis. And some sources have suggested that film attendance is down because countries have failed to keep pace with cinema renovation and the construction of multiplex facilities (Werba, 1988).

Nonetheless, one of the major reasons for box office decline cited in the literature is the rise in popularity of home video. And why not? Cinema ticket prices around the world experience annual increases and families cannot afford to attend the cinema regularly. Even among teenagers, who make up the largest group of cinema-goers, the VCR is popular. In a March/April 1988 study of VCR use in 986 VCR households in the United States, AGB Research found that teens watched more prerecorded material than adults (3.0 hours per week as opposed to 2.0 hours for women and 1.6 hours for men), while the preteenage group watched the next greatest amount (2.8 hours per week) (AGB Research, 1988).

In 1988 issues of *Variety*, the following 1987 one-year revenue declines were noted: 18.2% in France ("French admissions," 1988), 12.7% in Italy ("Italo attendance," 1988), 11% in Japan ("Japan biz fell," 1988), 16% in Israel (Fainaru, 1988), 3.5% in Australia ("Oz B.O. statistics," 1988), and 5% in Austria ("Austria's ORF chief," 1988). Arab countries are reporting a 50% decline over the last 10 years and directly fault the availability of the 5,000 video titles (Thomas, 1988).

Those same Arab countries are sites of high levels of video piracy. And where rampant piracy exists, cinema box office may be hurt even more, since first-run film releases often appear in the video shops before being screened at neighborhood theaters. Reports from Argentina, Brazil, Japan, India, and the Philippines describe the decline of box office receipts in those countries where video piracy is not under control (Cabera, 1988; "Blame homevid for sharp drop," 1987; Giron, 1987; Sen, 1984; Teixeira, 1987).

Investment in Film and Program Production. Another sign that video has had an international economic impact is the increased interest in domestic film and television production. Following the purchase of VCRs, people seek viewing content, and if that content is not available from the domestic media, they will look to imported content, usually in the form of feature-length films. Again, the rise in video popularity is not the only cause for this interest. The privatization of broadcasting in Europe and elsewhere, and the opening up of satellite and cable channels has also had considerable effect on expanding domestic demand for content. But the increase in VCR penetration has contributed to this demand. David Waterman predicts that new technologies and privatization policies in film and television producing countries that are U.S. trading partners will bring about greater production of domestic films and television programs in those countries (1988).

There is no way to tell what impact these products will have on international prerecorded video sales, but if the films and programs are popular in European cinemas and on television, they will no doubt be distributed on video too. It will be interesting to see how widely they will be diffused in the U.S. market, which has been relatively inhospitable to foreign products on television, at the cinema or on video cassette.

Over the last 20 years, film imports to the United States have declined considerably. From 32 countries the United States imported a total of 328 films in 1967, dropping to 186 films in 1987, for a 43% decline. Some of the largest drops were from France (56 to 23); the United Kingdom (79 to 36); Italy (65 to 15); and Spain (25 to 5) ("U.S. Film Imports," 1988). Foreign films on video have probably also experienced similar problems in the United States. Although data are not available, a quick glance at a catalog of available films on video in the United States reveals a very small section devoted to foreign products, and one is usually hard pressed to find more than a rack or two of foreign films in any video shop.

PIRACY: A SERIOUS ECONOMIC PROBLEM

If the United States dominates the world in its production and sale of prerecorded video, it only profits on this sale in parts of the world where video piracy is under control. Nowhere is the problem completely wiped out. In the United States, piracy accounts for about 10% of the market. It is a crime that is relatively easy to get away with, since all that is required is the ownership of two VCRs—one to play the original tape and one to record that content. And pirates rationalize that no one is hurt by this crime. Only the coffers of the large major U.S. film producers are affected in any degree, it is said. Calling video piracy the "'toxic waste' of the U.S. film industry," Jack Valenti, president of the Moton Picture Association of America (MPAA), claims film companies lose about $1 billion in sales to piracy annually ("Government claims U.S. is losing," 1988).

In a way, one could view this problem as media imperialism turned on its head, or as I have called it, "cultural imperialism by invitation," where media consumers, most of whom live in Third World countries, get to view a full range of U.S. films and television programs, and the copyright owners get no profit from the viewer's experience (Ogan, 1985a). The MPAA is trying to put a stop to international piracy, and spends about $15 million annually to support a technical staff of 400 people around the globe to combat the practice ("Jolly rogers," 1987).

Any country can fall prey to video pirates, but Third World and East European countries have experienced greater problems than has the West. A combination of factors, about which I previously written, contributes to this situation (Ogan, 1988a). The conditions that give rise to piracy are as follows:

> An environment with low media diversity.
> An environment that establishes trade barriers through high tariffs and tax rates or copyright fees for an imported product.
> An environment of minimal penalties for violation of domestic or international copyright law.
> An environment that permits use of accounting practices that mask the scope of illegal activity.
> An environment of scarcities in an inflationary setting.
> An environment with high freedom of entry to the black market system.
> (pp. 166-68)

More of these conditions are found in the Third World and in East Europe than in the West, although a high piracy level was the norm in Europe prior to the opening up of new media channels and the tightening of controls on copyright violations.

Since video piracy does considerable economic harm to the more fledgling film industries in the Third World countries, and since the losses in tax revenues from piracy to governments in those countries are significant, the control of pirate activity is very important to Third World economies. For example, India's profits from film exports were cut in half over a five-year period following 1980-81 when they peaked at about $15,000,000 (Cunha, 1988), and this drop has been attributed to the piracy of video.

In the same article in which I describe conditions that provide a climate for piracy, I suggest the following measures, which if taken in combination, should make effective policy for combating the problem.

> 1. Joint ventures with other film-producing countries and subsidies to the domestic film industry to expand diversity of local media offerings.
>
> 2. Use of nonprohibitive tariffs on VCRs and on videocassettes and reduction of cost of prerecorded content to a level that competes effectively with pirated products.
>
> 3. Use of technical measures to detect and trace pirated products to their origins.
>
> 4. A combination of legal measures that specifies penalties for violation of copyright of videotaped materials, makes those penalties stiff enough to outweigh the risk of breaking the law, and rewards police and customs officials to enforce the law.

5. Application of political pressure to clean up black market operations by aid-giving governments, international intellectual property organizations and domestic and international film producers. (pp. 170-176)

These measures were suggested by descriptions of actual cases and by economists who have specified policy for other black market activities. But no actual testing of either the conditions that give rise to a climate for piracy nor effective measures for reducing levels of video piracy in a country had ever been conducted. Testing is difficult because the measures of such variables as media diversity, level of piracy, effectiveness of legislative measures, and so on, all contain degrees of inaccuracy and many are merely "guesstimates" (see Table 13.1 for an example of differences in VCR penetration figures by source).

For example, for nearly every country where piracy of video products exists, a percentage estimate of the piracy level is specified, usually anywhere from 10% to 100%. If an exact percentage could be determined, on what basis would it be calculated? Here are some possibilities. Piracy levels could be based on the percentage of illegal film and television titles on the market, the percentage of number of rentals of any illegal title, the percentage of the revenue obtained from illegal content sale and rental, or the percentage of video shops that deal in illegal products. If one assumes that accurate data on any one of these measures could be obtained, how then could one be sure that every country bases its piracy levels on the same measure?

Most of the variables related to the conditions that give rise to piracy and the strategies I offer for effectively reducing piracy are subject to some of the same measurement problems. Availability of primary data on a country-by-country basis is also a problem in trying to test empirically the success of the identified strategies. With these fairly serious problems in mind, I undertook a study of 43 countries to assess the validity of these notions of international pirate activity.

The research was conducted because if the conditions that give rise to piracy and others that worked to control it could be predicted policies for countries to adopt for more successful control of the problem could then be proposed.

Methodology. The data on the 43 countries for the study[3] were collected from a variety of sources. Some statistics came from World Bank, United Nations, or UNESCO sets of indicators. An unpublished UNESCO-sponsored study (Alvarado, "Final report on the international flow," 1986), the *World Radio-TV Handbook*, the Motion Picture Association of America, *Variety*, *TV World*, *Communication Research*

Trends, Radio Liberty Research, Intermedia, Europa Yearbook, and some other trade publications were sources of most of the remaining data. In spite of the accuracy problems mentioned above and of the problems with using any data published in secondary sources, these sources constituted the only data available. No claims are made on the absolute accuracy of the data, but several of the variables (such as VCR and other mass media penetration) were checked through use of multiple sources. Where estimates were used from trade journals or other sources, it is believed that such estimates are as good as any data available.

Hypothesis for Study. Drawing on the above cited list of conditions that provided a receptive environment for video piracy, the following hypotheses were posed:

1. The less media diversity in a country, the higher the rate of video piracy.
2. The weaker the legal infrastructure and the penalties for copyright violation, the higher the rate of video piracy.
3. The less accessible prerecorded video content is in a market, the higher the rate of video piracy. This hypothesis can also be extended to availability of the videocassette recorders too, since black market activity in one area might stimulate use of other black market sources. In some countries of East Europe, few legal VCRs are available, but a thriving black market exists.
4. The weaker the economic conditions in a country, the higher the rate of video piracy.

These four hypotheses were tested separately and in combination. Multiple measures of the concepts listed in the hypotheses were used, as follows:

Media Diversity. This concept was operationalized in several ways. Media diversity was first measured by the number of media outlets available—the number of television channels, radio stations, daily newspapers, and cinemas in the country. It was also measured by the number of hours in the television broadcast programming day. Consideration of the amount of domestic content produced was made by examining the percentage of locally produced versus imported television programs and films. Finally, membership in the European Broadcasting Union was considered, since such membership brings with it increased opportunity for broadcast variety.

Legal Infrastructure and Penalties for Copyright Violation. Since many countries have had to revise copyright laws to accommodate videotaped content, one measure of these concepts was the existence of law passed within the last five years. The length of the jail sentence and the maximum amount of fine assessed for violation of the law was also measured. Whether or not a country had membership in one of the international copyright conventions—Berne or the Universal Copyright Convention—and whether a local organization had been developed to fight the problem, were also measured.

Accessibility of VCRs and Videocassettes. This was measured through the mean price of a locally obtained VCR, the rental price of prerecorded videocassettes, whether or not a country produced VCRs cassettes domestically, the size of the window (amount of time) between screening of a film at the cinema and its availability on video cassette, and number of video rental/sale shops in the country.

Economic conditions were measured through per capita GNP and international per capita debt. The debt variable was included because if a country had a high amount of foreign debt, availability of foreign currency might also be a problem and the country might be less inclined to enforce copyright, since it would require foreign currency for payment.

Video Piracy Rate was measured in percent at two time points—1985 and 1987, but the 1987 levels are the ones used in the hypothesis testing, unless otherwise noted.

FINDINGS

When the hypotheses were tested in combination through multiple regression analysis, they were not totally supported. This can mostly be explained by the size of the sample, 48, which was not sufficient to test the simultaneous relationship between so many variables. Multicollinearity among the independent variables was a problem that could not be corrected by increasing the sample size. Complete data were not available for additional countries. In fact, four countries that were originally included had to be dropped for lack of information, reducing the total sample to 43 (see note 3 for the complete list of countries). The problem became that developed countries in the West formed one group and developing countries formed a second group. With so few countries and even fewer cases of Third World countries that have been successful in significantly reducing their piracy rate, it was difficult to test the hypotheses in a regression equation.

Hypothesis 1 was partially supported. A higher piracy rate was found in countries with low numbers of television sets per capita ($r = -.49$; $p=.003$); low numbers of cinema seats per capita ($-.65$; $p = .000$); low numbers of newspapers per capita ($r = -.69$; $p = .000$) and low numbers of radios per capita ($r = .-34$; $p = .04$). (All piracy levels are for 1987 except where noted.)

Piracy rates were not significantly related to the number of TV channels, but were negatively related to the length of the broadcasting day ($r = -.35$; $p = .05$), or in others words, the shorter the television broadcasting day (and thus the fewer programs offered), the higher the piracy levels. High piracy levels were not significantly related to the number of domestic films produced per year, but were negatively related to the number of films exported to the United States from the country ($r = -.49$; $p = .004$), or the more films exported, the lower the piracy level. Piracy level was also negatively related to the number of U.S. films imported ($r = -.66$; $p = .000$), or the more films imported, the lower the piracy levels. Lower levels of piracy were also found in countries that were European Broadcasting Union members (Kendall's Tau $C = .36$; $p = .04$).

Hypothesis 2 was also partially supported. Membership in an international copyright convention was negatively related to high piracy levels (Kendall's Tau $C = .36$; $p = .04$) The presence of a local anti-piracy organization was not related to piracy level. Neither the amount of the fine nor the length of the jail term for copyright violation was related to the level of piracy. Since only 15 of the 43 countries had available information on the fines assessed and only 17 countries had information on the jail term, it is hard to interpret this relationship. Although not significant, both relationships were in the predicted direction.

Hypothesis 3 was also partially supported. The price of a VCR was unrelated to the level of piracy, probably because in countries where diversity is low, people have been willing to pay very high prices to increase viewing options. This may also explain the lack of relationship between piracy level and the presence of domestic production of VCRs. When it comes to availability of content, however, the hypothesis was supported. The rental price of prerecorded video was negatively related to piracy level ($r = -.39$; $p = .038$) and the number of rental shops per thousand population was also negatively related to piracy level ($r = -.35$; $p = .043$). Information on the length of time between cinema and video distribution was available for only 9 of the 43 countries, so it was impossible to obtain a statistically significant relationship with piracy level, although the relationship was in the predicted direction.

Hypothesis 4 was also partially supported. The level of piracy was negatively related to the GNP per capita ($r = -.43; p = .01$). The piracy level was positively related to the amount of external debt per capita in 1985 ($r = .41; p = .02$), but that relationship was not significant for 1987.

This study took more than a year in the gathering of data, so it is disappointing that I could not determine which of the independent variables predict increased or decreased piracy in the various countries. The presence of so many positive correlations is heartening. But it still cannot be said that the model is supported by empirical evidence, since western industrial countries tend to be those with the most effective legal system for prosecution of pirates, have the greatest economic resources a bring to bear on the problem, and are also the locations of the most media diversity. That is not to say that Third World countries have not or cannot follow the outline of this model to reduce piracy levels. I have studied the case of Turkey, for example, in developing this model. Turkey has been quite successful in reducing a nearly 100% pirate market to one with a much lower piracy level (Ogan, 1988a). MPAA officials have described their success in reducing piracy in several countries where they have helped establish local antipiracy organizations (Nix, 1986).

CULTURAL IMPACT OF VIDEO

It is clear from the preceding discussion that video has had a major economic impact, both positively for film- and television-producing companies that market products on video tape, and negatively for those same companies that lose profits to pirates who market those products illegally.

The cultural impact of those products on consumers around the around the world is much more difficult to determine. Since VCR owners, and not governments or broadcast networks, have control over the choice of content and the time of viewing, it is difficult to know what is being selected by those viewers. Even in the United States, where television viewing behavior is measured in fairly sophisticated ways, it has been difficult to determine what video content is being viewed (Kipps, 1987).

The widespread practice of video piracy makes the control of content distribution and the determination of content selection much more difficult. If a government that has serious reservations about importing cultural products finds itself unable to control the domestic black market in video, it might as well forget about controlling imports of television programs and films to be shown in domestic cinemas. For as VCR penetration increases, it is not confined to urban areas, but spreads to the most remote villages in developing countries.

The corresponding spread of western entertainment to those remote spots has concerned critical scholars as well as Third World policy makers. It has been argued that western economic, political, and social values are transferred to the Third World via films and television programs, and the VCR makes that transfer more pervasive (see Ogan, 1988b for a discussion of this argument). Some countries, such as those in the Arab Gulf region, tried to limit access to particular kinds of material to their citizens by enacting laws and/or guidelines about who can own a video shop, what kinds of material can be rented and who is able to rent the material (see Boyd, 1987; Ganley & Ganley, 1987).

But the evidence for the notion that western values are transferred through video-taped content has mostly been based on a calculation of the percentage ratio of imported to domestic television programs and films in a country. Research on what people actually watch and how they are affected by this content has been minimal (see Ogan, 1988b; Tracey, 1988; and Tracey, 1985). Following an analysis of a number of secondary sources, Tracey has concluded that television audiences do discriminate and their preference is for domestic programming over imported content (Tracey, 1988).

Another way of measuring the cultural impact of content has been through textual analyses of scripts. In this approach, it is assumed that if the encoded content promotes western consumerist values, such values will be transferred to the viewer of that content. But as Wren-Lewis points out in his criticism of David Morley's work on encoding/textual analysis, this is an "overtheorized area," on which abundant research has been conducted, while audience decoding is an undertheorized area that lacks empirical attention (1983, p. 195).

Mike Featherstone, a researcher of international cultural issues, is also critical of the failure to examine audiences. Featherstone would like scholars to investigate the range of responses to consumer goods and images in non-western societies and relate them to the interests and activities of different classes and social groups. He argues that the responses to images (such as those transmitted via western video content) may be interpreted differently by media consumers as "the good life by national elites, integrated and combined with elements of tradition by others, lived out in practice while rejected by yet others" (1987,

p. 31). Featherstone believes that what people do with media messages that promote consumerism is a more important issue to research than what the ideological basis of the messages might be. And since consumer cultures encourage people to exercise individual taste in their selection of goods, "massified" taste will never result from viewing such messages (1987).

Featherstone draws on the work of Pierre Bourdieu, a French sociologist, to illustrate that point. Bourdieu conducted a survey of more than 1,000 French people in 1963 and 1967-68. In general, Bourdieu attempted to determine the relationship between education and social origin and cultural knowledge and preferences, finding that taste was highly differentiated even within certain classes and educational levels (1984). Bourdieu was studying the French people and found significant intracultural variations. Examination of this issue across culturally should reveal much wider variation.

My own work in Turkey provides more evidence that tastes are differentiated by class, educational background, orientation to foreign cultures and languages (Ogan, 1988b). Turkey is a country of relatively homogeneous cultural background. The population is 85% Turkish speaking and 98% Muslim. Although most of the people share a common language and religion, their education, economic status, and lifestyle vary widely. You can still find people who dress traditionally, live in one-room structures without electricity and running water and who speak such thick Turkish dialects that it is difficult for an urban resident to understand their speech. At the other extreme are the millionaires who have luxurious winter homes in Istanbul on the Bosphorous and summer homes complete with private yachts on the Mediterranean, who speak English and perhaps other languages fluently and who mix in international circles as comfortably as they do at home. A growing middle class works in government service or private business, struggles to get its children through private schools where English or French is the language of instruction and on to one of the universities where the applicants are many and access is difficult. A people with such varied circumstances and life-styles would not be expected to share the same tastes in entertainment content, and my exploratory research on the selection of video-taped content confirmed that.

Until 1986 Turkey had only one national TV channel with a limited program schedule. A second channel was added in October of that year. The film industry, which at its peak produced 208 films in 1976, has been experiencing economic problems in recent years. Only 65 films were produced in 1981, and and from 1980-85, an average of 50 films per year were produced ("Shakeup in Illicit," 1987). The drop in the number of films produced can be partially attributed to a rise in the popularity of the video cassette recorder. The last estimate of Turkish VCR penetration placed the number at 2.5 million in 1986[4] (Welford, 1986). In addition to the range of Turkish films available, as many as 4,000 foreign films and television programs could be rented from video shops around the country (Welford, 1986).

I view the VCR as a liberating technology that allows viewers a range of taste in media content never before possible in Turkey, as well as other Third World countries. When mass media delivered the entertainment or information, the content choice was made by the distributors. Viewer choice—of television programs and films at the cinema—consisted of watching what was offered or rejecting it. Often only one channel was available, as in Turkey, and only one neighborhood theater was accessible. The diffusion of the VCR has brought with it a wide range of content—both foreign and domestically produced.

In order to research the cultural impact of that content, the first point to examine is the choices made by viewers. By interviewing video shop owners and taking a general inventory of stock in Ankara, Turkey neighborhoods, I was able to determine, much like Bourdieu, that the choice varies by the socioeconomic status of the viewer, and probably by the educational level and foreign language competence. That relationship might look like Figure 13.1.

What this graph indirectly demonstrates is the relationship between the cultural values of the media consumers and the values of the cultural products. When viewers could not make choices of media content, they were held captive. If there was no match between the values of the viewers and those of the film or television program, viewers could do one of three things: (1) They could decide not to watch what was offered (selectively attend); (2) they could reinterpret the content to fit their own value structure (selectively perceive); or (3) they could change their own values to match those of the media product (undergo attitude change).

In a panel study of Turkish residents in a squatter settlement outside the capital city of Ankara over a 23-year period, I discovered that as respondents aged, they paid less attention to television and films (particularly imported ones), frequently citing their failure to understand the content as the reason for not watching or attending[5] (Ogan, 1987). Respondents even specified that they watched certain programs on television if they were Turkish in origin, but did not attend to foreign programs or foreign music on the radio. Though people in this squatter

Proposed relationship between orientation to foreign culture and preference for foreign media products

	Level of income; education; proficiency in foreign language; number of trips, length of stay abroad		
	Low	Medium	High
Program/film choice			
Foreign films (not dubbed or subtitled)			
Subtitled foreign films			
Domestic films (ranging from traditional to Western themes)			

Figure 13.1. Proposed Relationship Between Orientation to Foreign Culture and Preference for Foreign Media Products

settlement didn't have the financial means to purchase VCRs, the local coffeehouse owner rented a different film every night for customer viewing. He said he only chose Turkish films with religious themes or Turkish comedies for his customers. The only foreign films selected were U.S.-made cowboy films. Squatter residents are relatively homogeneous in their educational and economic status, with few of them having more than a fifth-grade education, and most of them espousing conservative moral positions. So it was not surprising that this group of people expressed concern over the immoral content of imported and domestic films and the nudity in magazines and newspapers, and that they also objected to western popular music on radio. In other words, when television or the local cinema didn't offer a choice of content, these people chose not to watch or listen to objectionable material.

Another study of 294 VCR owners and their families, and of 125 male viewers of cassettes in bars and coffeehouses in Eskisehir, Turkey, found that choice of viewing content was related to the income, education and age of the respondents (Leek & McIsaac, 1987). Less educated respondents (who had not attended high school), expressed interest in Turkish films, while those with more education did not list domestic films among their choices. Of VCR owners and their children, more than half said they watched Turkish comedy on cassettes. Though this research didn't explore the issue of cultural origin of content and viewer choice, it demonstrates that when given a choice, not everyone selects foreign over domestic content, and that choice varies by cultural orientation that may be partially based on social class and educational background.

The Turkish studies support Bourdieu's research findings in the area of cinema attendance (1984). His survey of French respondents found that attending the cinema was a function of education, income, and age. He also found that Parisians were more likely to attend the cinema than provincial residents (p. 26). When Bourdieu examined film choices of viewers, he found that high brow films were the top choices of secondary school teachers, and such films were lower on the preference scale of professionals and lowest on the list provided by industrial and commercial workers.

DIRECTION FOR RESEARCH AND THEORY DEVELOPMENT

Obviously I am not the first to suggest there are flaws in the research findings and the theoretical development in the areas of media effects and uses and gratifications research. Critics have long held that findings produced in an experimental situation where the researcher selects the content and the viewer merely reacts to it in self-report of attitudes and/or observations of behavior change do not necessarily reflect real world situations where viewers select content and react to their own media choices. And I am also not the first to point out that effects research has not been very well developed cross-culturally.

Katz is one of the few scholars who have begun to explore audience reaction to U. S. television programs, namely "Dallas," cross-culturally, through focus group discussions with small groups of viewers. His research findings lead him to hold the position that television program meaning is not uniquivocally imposed on passive viewers, but rather such meaning is interpreted by viewers in terms of their cultural orientation (1984, pp. 28, 32).

Through Wren-Lewis suggests conducting more audience research for increased understanding of the decoding process, he thinks that such focus group interviews are not the best way to do it. Rather, he would conduct individual interviews and combine the results of such study with a textual analysis of the range of possible "readings" the text allows. He thinks a more complete understanding of decoding can be reached this way (1983, pp. 195-96).

Since Gunter and Levy have found that video viewing is a more isolated experience than television viewing (1987), it is probably even more important that individual interviewing take place when trying to determine the cultural impact of video content. The problem, of course, is in the self-report method. It will be important in such research to probe reactions in depth and shortly after viewing of particular content.

Theoretically, the research of decoding behavior fits best under the more general concept of audience activity. Levy and Windahl (1985) have explicated this concept, defining it as a "voluntaristic and selective orientation by audiences toward the communication process" (pp. 109, 110), noting that acknowledgment of activity is central to the uses and gratifications tradition. The authors point out that even in the structural/cultural framework of communication research in which the social regulation of media content and exposure behavior is assumed, there is variation in audience activity (pp. 111, 112).

Levy and Windahl were writing about audience activity as it relates to the mass media in that article. The VCR allows the audience much greater latitude for activity. In an article on the social significance of the VCR, Noble views the VCR as "guerrilla television" (1988, p. 139). It is being used, argues Noble (as well as Ganley & Ganley, 1988), to express a wide range of political views and to provide material that is specifically designed for a particular target group. And that is not only happening in the United States with the likes of Jane Fonda's workout tapes. Video-taped content ranges from instructions in assassination to motivational messages for terrorists to alternative reporting on an event from that of a government-owned broadcast station's version, to political campaign messages designed for special interest groups.

The introduction of new content and new opportunities for viewer choice makes it possible, as it never was before (especially in developing countries), to study the role of popular media culture intraculturally and cross-culturally. Where there is more domestic content available, as in countries with developed film industries, there will be opportunity for wider variation and a more accurate assessment of cultural preferences. It is important that we get going on this work before we make any more assumptions about the impact of the western capitalistic value system on the international population. Otherwise, the current era will be missing a history of its popular cultural orientation, much like the previous eras have. As Oscar Handlin wrote in his "Comments on mass and popular culture" (1959), it has been difficult to determine the effect of mass culture on popular culture because of a lack of a canon and a history of popular culture.

We have a record of U.S. mass culture and much of it can be examined at the Museum of Broadcasting in New York. We have no such record of mass culture in the Third World. And we know less about its development and impact. And while we might be able to observe the televised or filmed cultural products of developing countries, we have extremely little information about the emerging videotaped content in those countries. We know little and have assumed much in the area of international mass culture. It's time to test our assumptions.

NOTES

1. The author thanks Jian-hua Zhu for his assistance in the research and his contribution to the substance of this chapter.

2. For a discussion of the VCR industry see Lardner, J. (1987). *Fast forward: Hollywood, the Japanese, and the VCR wars.* New York: W. W. Norton & Company.

3. The countries analyzed in this study were Argentina, Australia, Brazil, Canada, Chile, China, Colombia, Egypt, Finland, France, Greece, Hong Kong, Hungary, India, Ireland, Israel, Italy, Japan, Jordan, Kenya, Kuwait, Lebanon, Malaysia, Mexico, Netherlands, Norway, Peru, Philippines, Poland, Portugal, South Korea, Saudi Arabia, Spain, Sweden, Taiwan, Thailand, Turkey, United Kingdom, United States of America, Union of Soviet Socialist Republics, Venezuela, West Germany, and Yugoslavia. Belize, Botswana, the Ivory Coast, Nigeria and Zaire had to be dropped for insufficient data.

4. The 1988 statistics appearing in the appendix do not reflect any increase in that estimate, another reason to weigh estimates with caution.

5. The original study was conducted in 1962 by Granville Sewell. It was his doctoral dissertation. Sewell shared the original data and the names and addresses of respondents with me, and I returned to Aktepe in 1975 and again in 1985 to survey the same respondents. Of course there was considerable sample mortality over the time period, but many of the original respondents remained in the community over the entire period. See Sewell, G. (1964). *Squatter settlements in Turkey: Analysis of a social, political and economic problem.* Cambridge: Massachusetts Institute of Technology.

REFERENCES

AGB Television Research. (1988). *VCR's: Changing viewing in American homes.* New York: AGB Television Research.

Alvaredo, M. (Ed.). (1986, April). *Final report on the flow of video hardware and software*. London: Broadcasting Research Unit, submitted to UNESCO, Paris.

Austria's ORF chief says lay blame for B.O. drop elsewhere. (1988, April 27). *Variety*, p. 200.

Barakan, M., et al. (1985). *The examination of video as mass media in Turkish society*. Eskisehir, Turkey: Anadolu Universitesi Basimevi.

Bourdieu, P. (1984). *Distinction: A social critique of the judgement of taste*. Cambridge, MA: Harvard University Press.

Boyd, D. (1988, May). *The videocassette recorder and the dissemination of western cultural and political information in the U.S.S.R. and Soviet-bloc countries*. Paper presented at the International Communication Association convention, New Orleans.

Boyd, D. (1987, May/June). Home video diffusion and utilization in Arabian Gulf states. *American Behavioral Scientist, 30*(5), 544-555.

Cabera, E. (1987, October). South on Latin America: prizes but no profits. *South*, p. 10.

Cunha, U. (1988, May 4). After a 3-year dry spell, India back in fest lineup. *Variety*, p. 302.

Fainaru, E. (1988, April 6). Israeli attendance fell 16% in 1987, but exhibs push on. *Variety*, p. 29.

Featherstone, M. (1987). Consumer culture, symbolic power and universalism. In G. Stauth & S. Zubaida (Eds.), *Mass culture, popular culture, and social life in the Middle East* (pp. 17-46). Boulder, CO: Westview.

French admissions dropped off 18.2%. (1988, March 30). *Variety*, p. 39

Gallup poll shows VCR penetration's more than thought. (1988, July 6). *Variety*, p. 35.

Ganley, G., & Ganley, O. (1987). *Global political fallout: The VCR's first decade*. Norwood, NJ: Ablex.

Ganley, G., & Ganley, O. (1988). The political implications of videocassette recording. In B. Compaine, (Ed.), *Issues in new information technology*. (pp. 265-291). Norwood, NJ: Ablex.

Giron, M. V. (1987, February 25). Filipino filmmakers having fits; piracy problem out of control. *Variety*, pp. 5, 462.

Government claims U.S. is losing $23-bil revenue to piracy globally. (1988, March 16). *Variety*, p. 76.

Gunter, B., & Levy, M. (1987, May/June). Social contexts of video use. *American Behavioral Scientist, 30*(5), 486-494.

Handlin, O. (1959). Comments on mass and popular culture. In N. Jacobs, (Ed.), *Culture for the millions* (pp. 63-70), Princeton, NJ: Van Nostrand.

Italo attendance fell 13.2% in February. (1988, March 23), *Variety*, p. 39.

Japan biz fell sharply in '87; Yanks hurt, local distribs fair. (1988, March 23). *Variety*, p. 39.

Jolly rogers flying high. (1987, October 31). *Economist*, p. 65.

Katz, E. (1984, May). Once upon a time in Dallas. *Intermedia, 12*(3), 65-81.

Kipps, C. (1987, October 14). It's hard to get a handle on HV data: A 'need to know' growing factor" *Variety*, pp. 145, 156.

Levy, M. (1980). Home video recorders: a user survey. *Journal of Communication 30*(4), 23-27.

Levy, M. (1983). The time-shifting use of home video recorders. *Journal of Broadcasting & Electronic Media 27*, 263-268.

Levy, M., & Windahl, S. (1985). The concept of audience activity. In K. E. Rosengren, L. A. Wenner, & P. Palmgreen, (Eds.), *Media gratifications research: Current perspectives* (pp. 109-122). Beverly Hills, CA: Sage.
Mid-year video review. (1988, June). *Screen Digest*, pp. 133-136.
Motion Picture Association of America. (1988, June 21). Estimates of Population, television household/sets, and VCRs. New York: MPAA.
Nix, W. (1986, December). Motion Picture Association of America. Personal interview.
Noble, G. (1988). The social significance of VCR technology: TV or not TV? *The Information Society, 5*, 133-146.
Ogan, C. (1985a). Cultural imperialism by invitation. *Media Development, 32*(1), 2-4.
Ogan, C. (1985b, March). Media diversity and communications policy. *Telecommunications Policy, 9*(1), 63-73.
Ogan, C. (1987). Mass media use factors in a Turkish squatter settlement: A 23-year panel study. *Gazette, 40*, 145-66.
Ogan, C. (1988a). Developing policy for eliminating international video. piracy. *Journal of Broadcasting & Electronic Media, 32*(2), 163-182.
Ogan, C. (1988b). Media imperialism and the videocassette recorder: the case of Turkey. *Journal of Communication, 32*(2), 93-106.
Oz, B. O. statistics. (1988, April 20). *Variety*, p. 56.
Sen, A. (1984, October 19). New laws aim at wiping out trade in pirated videotapes. *India Abroad*, p. iv.
Shakeup in illicit market due as Turkish homevid goes legit. (1987, February 25). *Variety*, pp. 354, 426.
Survey backs video erosion; rentals off, VCR resistance up. (1988, July 6). *Variety*, p. 35.
Teixeira, M. (1987, March/April). Pirate bounty in Brazil. *TV World*, pp. 35-36.
Thomas, M. (1988, May 25). Homevideo piracy, penetration killing Middle East cinema. *Variety*, pp. 1, 76.
Tracey, M. (1988, March). Popular culture and the economics of global television. *Intermedia, 16*(2), 9-25.
Tracey, M. (1985). The poisoned chalice? International television and the idea of dominance. *Daedalus 114*(4), 17-55.
U. S. film imports 1987 vs. 1967. (1988, May 4). *Variety*, pp. 203-223.
Waterman, D. (1988, June). World television trade: the economic effects of privatization and new technology. *Telecommunications Policy*, pp. 141-151.
Welford, R. (1986, April). Where Muslims meet Mike Hammer. *TV World*, p. 30.
Werba, H. (1988, May 4). Filmers eye TV competition warily. *Variety*, p. 311.
World video markets review. (1987, November). *Screen Digest*, pp. 249-253.
Wren-Lewis, J. (1983). Encoding/decoding model: Criticism and developments for research on decoding. *Media, Culture and Society, 5*, 179-197.

14

The Videocassette Recorder in the USSR and Soviet-Bloc Countries

DOUGLAS A. BOYD

This chapter explores recent communication technology in the USSR in light of contemporary political developments there. Its main focus is the home videocassette recorder (VCR), and specifically how the increased flow of video hardware and software to the USSR and Soviet-bloc countries may help expand the supply of information about the West.

SOVIET MEDIA, SOVIET INFORMATION

Since the first days of the Soviet Revolution, the Kremlin leadership has paid great attention to the propaganda value of information. Internationally, the USSR has been active in transplanting the Soviet media model to those states it either dominates or influences. In addition to being an early international radio broadcaster, the USSR realized that radio was the one medium that could be applied domestically to reach all areas of the vast country at one time. Radio had another obvious advantage to regimes for whom official *written* twentieth century political history could become an embarrassment.[1] By 1985, the USSR was transmitting more external hours than any other international radio service (*BBC annual report and handbook*, 1986). The Soviets increased the number of domestic radio and television channels and at the same time expanded the number of broadcast transmitters; in a country spanning 11 time zones, the USSR realized the value of using satellites as an alternative to terrestrial distribution of radio and television services. According to Soviet data, there are an estimated 93 million television receivers in this enormous country; in 1986, 93% of the population could receive one of the Moscow-based Central Television channels, and 82% could receive two channels (Lindsay, 1986).

It remains to be seen whether the Soviet Union has, in fact, entered a new age of international public diplomacy with the accession to power of Mikhail Gorbachev as General Secretary of the Communist Party. The casual observer notes some of the more obvious aspects of the new leader's influence: stylishly clad government press officers displaying keen insight about and knowledge of how to gain access to the Western press; a willingness to be more forthcoming in disclosures about such events as the Chernobyl nuclear disaster and the October 1986 fire and subsequent sinking of a Soviet nuclear submarine in the Atlantic. In December, 1986, Andrei Sakharov, the USSR's most famous dissident, was permitted to return to Moscow from internal exile in Gorky. On November 30, 1987, just one week prior to Gorbechev's December 1987 visit to Washington, D. C., NBC television broadcast an unrehearsed primetime one-on-one Tom Brokaw interview with Secretary Gorbachev.

There is an openness—*glasnost*[2]—campaign by Gorbachev and his associates to liberalize some of the traditionally harsh and tightly controlled aspects of Soviet life. On the other hand, the Nicholas Daniloff affair clearly signaled to the international press corps in Moscow that the new information policy did not necessarily apply to Moscow-based international journalists; foreign correspondents in Moscow were reminded by the incident that the free flow of information—at least as the term is used in the West—is confined to what the Party deems appropriate for release via official print and electronic media channels.

The Soviets jammed Ronald Reagan's Voice of America broadcast of his 1987 New Year message to the Soviet Union. The present leadership of the Soviet Union seems to have a very different information agenda from that prevailing in the West, despite the rhetoric surrounding *glasnost*. The Soviets are unlikely to lessen to any great extent their determination to uphold one of the pillars of Marxist-Leninist philosophy: control of information.

Home satellite reception and direct radio broadcasting from satellites are threats to Soviet control over incoming information about the outside world, specifically from the West. Additionally, individual use of computers, and hence the ability of private citizens to build data bases, holds future prospects for a freer communication flow to and within the USSR (Malik, 1984). Four years after Malik's pre-*glasnost* examination of computers in the Soviet Union, Dizard and Swensrud (1988) observed that, "One of the hallmarks [of the new openness] will be the modification and phasing out of long-standing controls over communications and information resources" (p. 10).

The VCR, however, is already operational. While estimates of the number of home video recorders in the Soviet Union vary and surely give an unreliable picture of actual VCR penetration, it seems clear that there are a number of machines operating in the homes of those with the money and position to purchase them. Faculty at the Moscow University School of Journalism place the number of VCRs in Moscow at 100,000 (faculty members, personal communication, School of Journalism, Moscow University, Moscow, USSR, January 3, 1988). A 1986 estimate placed the number of VCRs in the USSR at between 250,000 and 300,000 (Yasmann, 1986), a small number for this populous country. Video recorder ownership is largely an urban phenomenon; for example, a 1987 Radio Liberty report estimates the number of private VCRs in Leningrad to be 30,000 (Yasmann, 1987). In September, 1987, *Pravda*—apparently ignoring the existence of imported machines—estimated the number of VCRs in the Soviet Union to be 350,000 (Parfenov, 1987).

Most Soviet citizens who possess video recorders are the elite, people with funds to acquire one either from someone permitted to travel abroad, to purchase merchandise at a state hard currency store, or to buy from an underground contact, or from someone on a foreign military or diplomatic assignment. A VCR in the USSR is likely to serve more people than is the case in the West, where, as Gunter and Levy (1987) found in their British study, VCRs tend to be used by immediate family members and close friends. Finally, as in the early stage of VCR adoption in Asia and Africa, one's desire to obtain a VCR may be profit-motivated (Boyd, Straubhaar, & Lent, 1989). A citizen with a video machine at home can charge for showing tapes: pornography, sought-after western films, or illegally circulating Soviet material.

FREEDOM OF INFORMATION IN THE U.S.S.R.

The control of information is one of the three basic instruments of rule in the Soviet system, the second being the Party's monopoly on decision making in politics, the economy, and culture; and the third being the police. Information control is the weakest of the three. (Shanor, 1985)

To the outsider, jamming of radio broadcasts is possibly the most well known state-organized means of keeping outside information from Soviet citizens. Until the recent relaxation of jamming against the major western broadcasters, the USSR reportedly spent at least as much to jam incoming signals as to transmit transnationally its own programming. Now, Soviet citizens wishing to hear foreign radio may do so. The future possibility exists of selling uncensored western news magazines and newspapers in the Soviet Union, but for now, printed information from the West is difficult to acquire because of thorough border and postal checks. Those wishing to obtain uncensored information rely on audio tapes of foreign broadcasts and tapes made by dissidents within the Soviet Union. For a number of years in the 1970s technology ran ahead of the law. The law applied only to printed information—it said nothing about audio tape. The result was *magnitizdat*, clandestine tape publishing.

There is also a well-organized underground press—known as *samizdat* (self-publication)—circulating typed pages of news, commentary, and banned works. Authorities are concerned that increasing numbers of personal computers and photocopy machines will cause underground publications to proliferate. Referring to the government's strict control of photocopying machines, Roald Sagdeyev, director of the Soviet Space Research Institute, has said that scientific progress cannot be made "unless we stop treating copying machines like class enemies" (*The wizard of IKI*, 1987).

MOTIVATIONS FOR VCR OWNERSHIP

There are at least three motivations for Soviet citizens to acquire a VCR, to rent or purchase illegal cassette material, or to pay to view such material in private. First, VCRs are a means of making money. Second, they provide the best way to acquire forbidden fruit—western-produced feature films and even television news. Third, videocassette recorders are one means of circulating subversive material.

1. *To Gain Financially Via an Underground Business Activity*: The post-1917 underground economy in the Soviet Union has consistently been active in the same manner as such activities thrive in cultures where certain commodities are scarce and sought after and where those willing to risk the wrath of the authorities supply products in demand. In the Soviet Union, the government calls the sale of non-Soviet-manufactured machines and cassettes a black market function because home video is primarily an underground economic activity. The government first became concerned about VCRs in the early 1980s when official articles about specific illegal viewing parlors began appearing in the press to warn people about the dangers of such activity. Media reports told of places where people were charged a fee to see banned western films, on VCRs. An article in the Latvian newspaper *Sovetskaya Latviya* (Silinysh, 1985) told of people entering a former kindergarten room, paying a fee, and then "look[ing] through a secret keyhole into some incredible, cruel and corrupt world. A world of bloody passions, vices hyperbolized to the point of disgust, supermen and werewolves" (p. 141). Such pejorative and cautionary reports were officially sanctioned with the apparent intent of discouraging the consumption of western video material. Not only was the video underground fostering illegal economic activity the government had long attempted to stop, but possibly more damaging was the fact that VCRs were spreading information about the West, some of which was interpreted to be anti-Soviet.

2. *To Learn about the West from Western Productions*: People in both the Soviet Union and the United States are curious about the opposing superpower. Previously mentioned was the jamming of most western broadcasts in the major languages used in the Soviet Union, and the banning of most newspapers, magazines, academic journals, and films from abroad. This has caused a tremendous curiosity about the United States and western Europe that can be partially satisfied through feature films. There is apparently great entertainment value in viewing the recent U.S. films circulating on cassettes. The American visitor to the Soviet Union sees the lively activity among traders, enterprising young adults wishing either to purchase or to trade Soviet artifacts and products for used American blue jeans, brand-name running shoes, or T-shirts with American slogans on them.

3. *To Subvert the State*: Some in the Soviet Union see the possibility of using the VCR and other new technology as a means of fostering change in their country. Apparently a small minority of the population wish to change the political system radically. A somewhat larger number of citizens, although still a minority, see shortcomings of the political system that in their opinion need changing.

The VCR is the most recent addition to communication technology; short-wave radios tuned to western broadcasts, underground newspapers, audiocassettes of banned authors reading their works, and now VCRs all contribute to what Donald Shanor (1985) calls the *underground telegraph*. The Soviet Union has traditionally tried to stop the informal dissemination of any information that does not have official government approval. Sometimes warnings about such activities appear in the official press in an attempt to discourage consumption.

Just after referring to external "radio voices" (western medium- and short-wave broadcasts), V. M. Chebrikov, Chairman of the Soviet Committee for State Security (the KGB), noted in a published speech to the 27th CSPU Congress on February 28, 1986:

> A new problem connected with the widespread proliferation of home video equipment also merits attention. This phenomenon, which in itself is good and progressive, is being used by some people for the propaganda of ideas which are alien to us, the cult of cruelty, violence, and amorality. This activity must receive from the public a precise and uncompromising assessment. In our socialist society, just as on a well-tended field, there must be no weeds. ("Speech by Comrade V. M. Chebrikov," 1986, p. 12)

VCRS IN THE SOVIET UNION: THE BEGINNING

It all started not unlike the importation of some of the very early Sony, Beta, and VHS cassette machines into developing areas such as India and the Arab Gulf states. The prospect of showing video tapes of foreign television programming and feature films at home appeals to the higher-income groups in societies where there are limitations on the type and quantity of visual material available. VCRs are status symbols where they are either scarce or officially discouraged. With regard to the developing world, Boyd and Straubhaar (1985) note that dissatisfaction with programming on broadcast television and other forms of visual entertainment predispose people to purchase and utilize VCRs.

National and regional television channels in the Soviet Union serve a vast and ethnically diverse population. The visual medium, like every other form of mass communication, is programmed to promote the interests of the government. Referring to pre-*glasnost* television, Leonid Kravchenko (1988), noted that ". . . prime time current events shows were ostentatious and wordy in the reporting of industrial achievements. According to a sarcastic joke of the day the best way to fulfill the national food program was to hook your fridge to your TV set" (p. 2).

The USSR is a major motion picture producer, but with the state as both broadcaster and filmmaker, freedom of artistic expression is limited to what is officially tolerated. Although there have been some old U.S. television programs, such as *Daktari*, and *Mighty Mouse* cartoons on Soviet television, almost no other western-produced television programming and feature films are permitted. To open this "window to the West" would be at variance with one of the basic tenets of Soviet philosophy—strict control of both internal and external information. If it were technically possible to create the television equivalent of the Voice of America, Radio Liberty, or the BBC External Service, it would surely be jammed in a manner similar to the Soviet jamming of Radio Liberty broadcasts. The Soviet government has also suggested that it would destroy Direct Broadcast Satellites (DBS)—high-powered satellites designed to reach home viewers with small satellite dishes—operated by western countries with the aim of reaching Soviet citizens.

Home VCRs are relatively light-weight devices that, when attached to a television set, provide the means for a type of liberation from what is offered on broadcast television. Most developing countries did not realize that these seemingly neutral home appliances would potentially threaten regional film industries, such as those in Brazil, India, and Egypt, or seriously impinge on broadcast television. The Soviet government, however, realized the potential danger some time before the worldwide popularity of VCRs became apparent.

Initially, the Soviet government made no official comment on the number of VCRs smuggled into the Soviet Union and brought back by returning military and diplomatic personnel, but did remark publicly on the potential danger of machines themselves as well as on the illegal cassette material in use. In the spring of 1983, *Sovietskaya Rossiya*—the official newspaper of the Communist Party Central Committee—revealed that those in power were concerned about the illegally circulating western video material. Pointing to the dangers of watching such material, the paper told of a Soviet woman who had had to be confined to a mental hospital after viewing cassettes of western horror films. Additionally, the paper warned readers against black market gangs duplicating and distributing western films, giving *The Godfather*, *Clockwork Orange*, and *Straw Dogs* as examples (Smale, 1983).

A November 1982 law concerning items Soviet customs should inspect included VCRs because the machines—particularly those capable of showing tapes from the U.S. (NTSC) and the other European color system (PAL)—are able to show material that may contain "information that could harm the country's political or economic interests, state security, public order or the population's health or morals" (Schmemann, 1983, p. 13). In October 1983, the newspaper *Komsomolskaya Pravda* identified the illegal importation of VCRs and the copying and sale of western films as the Soviet underground's biggest growth industries, and the *New York Times* noted that, in addition to the three popular illegally circulated films mentioned previously, the most preferred contemporary American films on tape in the Soviet Union in 1983 included *The Deer Hunter*, *One Flew Over the Cuckoo's Nest*, and *Last Tango in Paris* (Schmemann, 1983). By 1985, Soviet authorities decided they should counter the spread of imported VCRs and stop the circulation of material deemed dangerous to the state.

THE GOVERNMENT'S VIDEO BUSINESS

In the early 1980s, the Soviet government signed a contract with a Japanese electronics firm to produce cassette recorders in the Elektronika factory in Voronezh, near Moscow, reasoning that when the video revolution arrived, the country should be in a position to produce its own hardware. However, the machines produced domestically were incompatible with the imported, mostly VHS, recorders from Japan. Moreover, the expensive Soviet-manufactured cassette recorders gained a reputation for unreliable, poor-quality reproduction, and repair difficulties. Those acquiring Soviet machines and desiring to view western-produced cassettes found that an underground industry had developed to convert the Soviet players for materials recorded on other systems (Taubman, 1985; "Videos, pirates," 1986).

To eliminate the thriving underground market in the resale of western VCRs, in early 1986 the Ministry of Trade mandated a 50% price reduction on imported electronic equipment in state-operated *komissionnyye* second-hand stores. This measure tried to bring items in these stores in line with prices in the *Beriozka* (hard currency) shops so as to make imported video equipment more available and thus weaken the video hardware speculator's position. However, the move had the opposite of its intended effect: those wanting to sell used imported video recorders simply started doing business with the private (underground) sector (Yasmann, 1986). Even those with the funds to purchase locally made VCRs had to pay the equivalent of 17 months' average wage and wait several months for delivery (Goldkorn, 1986). This writer found used imported machines readily available on the Moscow and Leningrad black market in exchange for hard currency. Imported VCR prices vary, depending on the age and the number of available features, but the price in early 1988 ranged from $1,500 to $3,500 (versus $250 to $600 for comparable machines in the U.S.). In the June 1988 issue of *Money*, Robin Micheli profiles the lives and finances of a married couple, both dentists, living on desirable Gorky Street in Moscow. The story mentions that among other consumer items, the couple owns a Hitachi VCR valued at $2,550 (Micheli, 1988). The most sought after imported machine is one capable of playing both European color standards: SECAM, the French-invented system used in the Soviet Union, and PAL, utilized by the television systems of western Europe, except France.

To combat the thriving black market trade in foreign-made machines, the government decreed that by the mid-1990s the quality and quantity of Soviet-produced VCRs must increase. Specifically, the stated VCR production goal of the Elektronika VM video recorder is 60,000 VCRs per year by 1990 to increase to 120,000 recorders by the year 2000 ("Videos, pirates," 1986). Increased numbers of Soviet-made machines will not necessarily solve the problem; as in the developing world, where imported VCRs are preferred over locally assembled units, Japanese and European brands will probably continue to be most in demand.

Visitors reaching the Soviet Union are likely to notice Soviet citizens carrying VCRs on the plane. Foreign-made machines may be imported as long as the owner pays 600 rubles [3] (around $300 at the official exchange rate) to customs authorities, and she or he can document that the foreign currency reserves of the state were not involved in the purchase. Most often, the foreign currency required to purchase a VCR comes either from family or friends abroad or from Soviet black market sources.

In an attempt to take the market away from the well-organized illegal videocassette underground,[4] the government began making Soviet films and approved west European productions—those not anti-Soviet, anti-Communist, or showing at least a balanced view of western society—available in video stores (Walker, 1986). Customers saw this activity, along with the state's decision to increase the production of VCRs, as a move by Secretary Gorbachev to decrease the technology gap between the Soviet Union and the West and to loosen the government's grip on the type of cassette material that could be viewed. Additionally, if successful, the decision could help stem the thriving underground market in recorders and western cassette programming. This would only be successful if consumers desired domestically produced machines and government-approved western tapes available through rental stores. Thus far, however, the Soviet government-operated cassette rental business has not been successful, despite its initial novelty value.

In the eleven video stores in Moscow, it costs between $2 and $4 to rent a cassette for 24 hours, and late return effectively doubles the price. Customers in a city where the average monthly wage is approximately $200 find this too much to pay. Much like video stores in the West, Soviet stores price new cassettes higher than old ones. In early 1988, this writer found that cassettes of old films and documentaries in Moscow's *Arabat* shopping area were much less expensive than newer films. The rental price of approved imported films from west European countries is three to four times that of old Soviet films. Some Moscow video stores show cassettes in small 50-chair viewing rooms, but they reportedly are not well patronized (Tinsley, 1986).

THE UNDERGROUND DUBBING AND DISTRIBUTION OF WESTERN FILMS

The video distribution network seems to work in a manner similar to other types of underground consumer businesses in the Soviet Union. Most commonly, a privileged person such as a diplomat, military official, or journalist brings a film into the country on cassette. The Soviet government suspects that western governments, possibly with the encouragement of their intelligence agencies, help inject selected feature films into the Soviet Union video underground via diplomatic mail privilege. Many western embassies in the USSR do, in fact, maintain a library of recent feature films on cassette for viewing by their own nationals and occasionally by local employees.

Once a film reaches the video underground, it is dubbed into Russian or another major Soviet language. This is accomplished easily, and often expertly and inexpensively, by first translating the dialogue into the appropriate language and then hiring actors and actresses to read parts on audio tape while the video is running. Some dubbings are crude, with a single voice narrating an entire western feature film, for example. The audio tape is then transferred to the audio track of a master cassette from which copies are made. Alternatively, dubbing is done directly onto a second video recorder as participants view the film on the master machine. Often actors and actresses will participate for no fee—their payment is the privilege of seeing the film. Thus, a dub can be made for the equivalent of under $40 (Taubman, 1985). This process may sound crude to readers familiar with professional dubbing, but the finished product reportedly is often quite polished and certainly acceptable to customers eager to see popular western films. In January 1988 this writer interviewed people in Moscow and Leningrad knowledgeable about U.S. films circulating on VHS cassettes. They reported that one of the most popular underground films, *The Untouchables,* was a single-voice dub into Russian.

UNDERGROUND VIDEO MATERIAL

From the government's viewpoint, the terms *violent* and *pornographic* tend to be interchangeable, while in the West both the legal and social definitions of the two terms differ. The Soviets have been worried about unapproved video material since the first VCRs were imported in the late 1970s. Making matters worse is the illegality of any kind of underground activity. Publicly, there are three concerns: pornography, unapproved western films, and uncensored foreign television news. However, from the govenment's viewpoint, the most dangerous information a Soviet citizen can receive is that which threatens the single-party supremacy. Specifically, this means information regarding religious rights, political rights, and ethnic rights—all of which the Soviet government thinks could foster dissent. Public anti-Soviet displays, such as those in Soviet Armenia in March 1988, are what concern the government most; actions are usually taken to at least slow information that could promote such activities.

Articles in the Soviet press in the early 1980s told of violent and pornographic video material being shown to both adults and children, detailing the alleged effects of such showings and the harsh punishments mandated by Soviet courts. So concerned were authorities about video pornography—known colloquially in the USSR as *Swedish tapes*—that a new criminal code promulgated in mid-1986 included a broadly defined video pornography section. The law and the punishments it entailed received a great deal of press attention, evidently with the intent of stemming the underground video market. One such article appeared in the February 28, 1987, edition of *Pravda*. Soviet Deputy Prosecutor General S. Shishkov discussed the law, told of instances where it applied, and gave examples of those arrested and punished for such activity:

> For example, a group of individuals who duplicated and sold video cassettes with particularly graphic scenes of violence, brutality, and pornography were arrested in Kharkov. Fully 15 sets of video equipment, nearly 400 films, and 185,000 rubles (approximately $220,000) in cash and valuables were confiscated from the accused during the investigation.
>
> Rakhmanov, a coach from the children's sports school in the city of Namangan, showed obscene films to his students in his apartment. Up to 20 adolescents attended, each one of them paying 10 rubles ($11.00) to see it. Gailyunas, a bartender from the city of Kaunas, showed similar films on numerous occasions to school pupils and musical school students in return for payment. (Poison from the screen, 1987, p. 6)

Similarly, in Perm, a city in the Urals, seven men were arrested and found guilty of watching a videocassette of *The Last Male Virgin in America*. All were fined and one was sentenced to prison and expelled from the Communist party (Branson, 1987).

These and other accounts in the Soviet press should not necessarily be interpreted to mean that the authorities aim primarily at pornography. The cities used in the examples and the ethnic names of those arrested seem chosen to alert readers that the danger is geographically widespread and affects all nationalities.

The authorities are also concerned that so much western material not permitted on Soviet television or allowed in theaters is reaching an elite population through VCRs. They worry particularly about popular western films that teach viewers violent and criminal acts. Films like

The Good, the Bad and the Ugly, *Who's Afraid of Virginia Woolf*, *Amadeus*, and *The Godfather*, as well as Bruce Lee movies, fall into this category, as do art films from European directors such as Ingemar Bergman, Federico Fellini, Milos Forman, and Bernardo Bertolucci. Banned also are videos of such rock music groups and singers as Duran, Duran and Boy George.

The video law deals with another type of underground videos: films promoting anti-Communism and anti-Sovietism. Here, according to Deputy Prosecutor Shishkov, the law applies to those propagating a "highly defined type of subversion directed against our country" (Poison from the screen, 1987, p. 6). Films such as *Rambo: First Blood Part II*, *Rocky IV*, *Moscow on the Hudson*, and *White Nights* are proscribed because they portray the Soviet Union or Communism in a bad light. There are underground Soviet-bloc-made films circulating on cassette. One, popular since 1985, is Polish director Andrzej Wajda's *Man of Iron*, a sympathetic examination of Gdansk and the birth of Solidarity (Taubman, 1985).

Finally, the illegal circulation of foreign television news probably worries Soviet authorities who may envision a day when home-made satellite receiving dishes will be so small that they can receive foreign television programming without having to be mounted—and thus visible—on roof tops. In the meantime, foreign television news, as well as U.S. television programming, reaches those wishing to see it through the Estonian capital of Tallinn. Because Estonian is similar to Finnish, many Estonians watch and enjoy the spillover from Finnish television. The Estonian Communist Party has said that Finnish television is a "channel for Reagan" and a U.S. "ideological weapon" (Rislakki, 1986). Dissidents regularly tape news and other programming from Finland and either translate or subtitle for speakers of the USSR languages ("Videos, pirates," 1986).

Portable light-weight video cameras have been acquired by dissident video enthusiasts in the Soviet Union. Authorities foresee that television—generally agreed in both the Soviet Union as in the West to be the most powerful medium—may become an outlet for domestically produced underground cassette material. It could become a force more persuasive, and thereby more threatening to the government, than illegal newspapers or audio recordings of foreign news. For example, video tape, shot by amateurs during the spring 1988 unrest in Soviet Armenia, circulates among dissident groups. Copies have been given to the Soviet bureaus of western television organizations; scenes of Armenian protests have even been used by the American networks.[5]

A particularly curious enigma of Soviet information control involves VCR technology. Official policy discourages the use of any unapproved western video material; yet an article in the January 1988 issue of *Radio*—a monthly technical magazine similar to *Popular Electronics* in the U.S.—provides step-by-step directions on how to modify SECAM-only VCRs to play PAL tapes (Ketners, 1988). A Soviet citizen's only motivation for making this conversion would be to view western cassette material in color.

VIDEO IN SOVIET-BLOC COUNTRIES

The VCR is becoming increasingly popular in the Soviet-bloc countries: Bulgaria, Czechoslovakia, East Germany, Hungary, Poland, and Romania. Historically, a few of these countries have exhibited some independence from the Soviet information model. Although the Soviet-style centralized approach to information-control is pervasive in the East European Communist countries, they have taken different positions on how both home video cassette recorders and programming for them should be handled. These countries have generally concluded that video cannot be stopped or, for that matter, effectively controlled. Like the Soviet Union, many East European countries manufacture their own machines together with socially and politically acceptable cassettes.

The Bulgarian goverment, for example, requires that VCRs imported or purchased in the country be registered with the state. All cassettes entering the country are screened and released only if acceptable material is on the tapes ("Moscow faces," 1986). The number of machines in the country is officially estimated to be 50,000 ("Special report," 1988).

Reportedly, in Czechoslovakia the government was initially reluctant to produce VCRs because it was thought that their spread would only exacerbate the underground video problem. Since 1985, machines have been assembled in the country from imported parts supplied by a subsidiary of the Philips company and a long-range plan for the development of a wide variety of home video equipment has been approved. It is estimated that there are approximately 150,000 VCRs in the country ("Special report," 1988).

As in most Soviet-bloc states, Czechoslovakia plans to produce material for home viewing. However, the government's worst fears have been realized; exiled Czechs in the West have been producing cassette material and smuggling it into the country for circulation. Of particular concern is the bi-monthly *Video Magazine* containing banned information such as that in support of the Solidarity union movement ("Videos, pirates," 1986).

The German Democratic Republic (GDR) has a governmental policy toward VCRs that is rather unusual for an East European country. Unlike most other states in this region, which have decided to live with and perhaps use VCRs to their own advantage, the GDR government is resisting them. East German authorities officially discourage the importation of machines by either visitors or citizens. The motivation for this appears to be financial rather than political. Most citizens of Soviet-bloc states do not have direct access to western television as do the East Germans, most of whom can pick up West German television networks (Boyd, 1983). They do not need a VCR in order to enjoy popular entertainment or news from the West—a necessity in other East European countries. However, by permitting the sale of VCRs in state-run hard currency stores and through the Genex service—allowing those in the West to purchase goods with hard currency for relatives in East Germany—the state gains financially from the sale of VCRs ("Videos, pirates," 1986).

Perhaps more than any other East European government, Hungary's has accepted the inevitability of video. In 1984 the Hungarian Mass Communication Research Centre surveyed 10,000 persons on their media habits. The study projected the results to estimate the number of privately owned VCRs at approximately 72,000, a number "surpassing even the boldest [previous] estimations" (Valko, 1985). By 1987, the estimated number had increased to 300,000 ("Special report," 1988). To supply programming for the home video market, the government has permitted a number of business cooperatives to function as production houses.

Poland has the largest number of VCRs in eastern Europe. In 1987 alone, the number of machines is estimated to have doubled; by the end of 1987, there were one million machines in the country ("Special report," 1988). Poland may be the birhtplace of East European revolutionary video. At first the state controlled video, but it soon became resigned to the fact that its spread is impossible to control.

The Catholic church has played an important part in the spread of VCRs in Poland because it has encouraged parish churches and halls to acquire machines. The state has started a chain of stores to distribute officially approved cassettes as well as western material ("Videos, pirates," 1986), but the most interesting and sought-after material comes from such directors as Andrzej Wajda or Ryszard Bugajski. *The Interrogation*, Bugajski's film about a female victim of Stalinism, enjoyed great underground video success (Goldkorn, 1986).

Data from Polish researchers provide an emerging picture of ownership patterns and usage. In 1987, about 32% of the Polish population watched video recordings; those viewing were more likely to be males between 15 and 29 years of age. VCR ownership is highest among the self-employed, those engaged in some form of private enterprise (Pomorski, 1988).

The Romanian government regards video and other aspects of state media as "instruments of the party," but it too has accepted the fact that video is so popular that it is impossible to control. The government has established a series of video clubs in resorts and youth clubs as a means of suppressing both the thriving black market for VCR hardware and the desire to view western-produced visual material (Video equipment, 1985). VCRs are not assembled in Romania.

A trickle of video software flows from East to West. For example, cassettes of programming pirated from Polish and Czechoslovak television can be rented in West German bookstores that cater to the East European emigree community.

Conclusion

To suggest that the VCR or other forms of communication technology alone will measurably change the Soviet Union and the East European countries defies both contemporary communication theory and the reality of Soviet-bloc political life. Yet it is interesting to speculate about the spread of other communication technologies—personal computers, photocopy machines, and satellite receiving dishes—and what effect they may have on the dissemination of information promoted by Soviet dissidents.

At least in the short run, western taped programming viewed on VCRs will not measurably alter the Soviet political view of the West. For example, witness the reality of East Germany, where television from West Germany has been available for over two decades. Although video is potentially a more powerful medium than radio, the consumption of taped material does not necessarily fill the need for news and information already satisfied by foreign radio broadcasts or the underground press.

Soviet authorities appear to be concerned about video's potential to cause further dissent, either via domestically made programming or that from outside. Uncensored western images from feature films and television programs in the homes of the Soviet elite could provide a detailed view of the West not available from other media. What that image is and what effect it will have in the long run is difficult to assess. At the very least, the home videocassette recorder, even in a limited number of Soviet and East European homes, opens the door to heretofore unavailable information. The home video recorder, like other new communication technology and the information it delivers, provides additional evidence that authoritarian states ultimately fail to dominate the minds of their citizens through censorship.

NOTES

1. An example is General Secretary Gorbachev's November 1987 speech via live radio and television on the occasion of the 70th anniversary of the Bolshevik Revolution. His address criticized Joseph Stalin for his "wholesale repressive measures and the acts of lawlessness. . . "(Keller, 1987, p. 1 Section 4). Gorbachev could refer to Stalin's "many thousands" of victims because the actual number—in the millions—has not been officially acknowledged by the Soviet government.

2. When the term first appeared in the Soviet press, the official news agency TASS translated it into English as propaganda. This was quickly changed to *openness* or *frankness*.

3. It is difficult to provide an American dollar equivalent of the ruble. A "soft" currency, it has no value outside of the USSR. The official exchange rate—the rate charged tourists—is, of course, very favorable to the state. In 1988, the official exchange rate was approximately one half dollar per ruble. However, one is openly approached in the Soviet Union by those wishing to change money at a rate as high as 6 rubles per dollar.

4. For example, there are persistent rumors that certain organized groups of people—astrophysicists and employees of the Gosteleradio and Orbita receiving stations—tape programming from western satellites and pass it to the underground video market. Whether true or not, the rumors give an interesting glimpse into the Soviet attitude toward foreign films and television programming (Yasmann, 1978).

5. Such scenes about Armenian protests were shown on the *CBS Evening News*, 6:00-6:30 p.m. (Eastern), Saturday, March 26, 1988.

REFERENCES

BBC annual report and handbook 1986. (1986). London: BBC.
Boyd, D. A. (1983). Broadcasting between the two Germanies. *Journalism Quarterly, 60*(2), 232-39.
Boyd, D. A., & Straubhaar, J. D. (1985). Developmental impact of the home video cassette recorder on third world countries. *Journal of Broadcasting and Electronic media, 29*(1), 5-21.
Boyd, D. A., Straubhaar J. D., & Lent, J. A. (1989). *Videocassette Recorders in the Third World.* White Plains, NY: Longman.
Branson, L. (1987, January 25). Blue movies furore makes Russia blush. *The Sunday Times* (London), p. 30.
Dizard, W., & Swensrud, S. B. (1988). USSR: Glasnost and the information revolution. *Intermedia, 16*(1), 10-19.
Goldkorn, W. (1986, November 23). Video dell'est. *L'Espresso*, pp. 71, 73, & 75. Translation undertaken by the United States Information Agency, Document No. 121844, PH/BL.
Gunter, B., & Levy, M. (1987, May/June). Social contexts of video use. *American Behavioral Scientist, 30,* 486-494.
Kravchenko, L. P. (1988, September). Paper presented at the annual meeting of the International Institute of Communications, Washington, DC.
Keller, B. (1987, November 8). Gorbachev speech: Playing to a wary public. *New York Times,* p. 1, Section 4.
Ketners, V. (1988, January). A PAL system signal decoder. *Radio,* pp. 27-29.
Lindsay, M. (1986, October). The coming of cable. *TV World,* p. 36.
Malik, R. (1984). Communism vs the computer: Can the Soviet Union survive information technology? *Intermedia, 12*(3), 10-23.
Micheli, R. (1988, June). The good life on 1,100 rubles a month. *Money,* pp. 48-56.
Moscow faces the new age. (1986, August 18). *Newsweek,* pp. 20-22.
Parfenov, L. (1987, September 25). Video: Present and future: The third screen. *Pravda* (2nd ed.), p. 3. Translated by Foreign Broadcast Information Service, FBIS-SOV-87-205, October 23, 1987, p. 25.
Poison from the screen. (1987, March 1). *Pravda,* p. 6. The original article appeared in the February 28, 1987, issue of *Pravda.* The author used the translation appearing in the March 1, 1987, English-language edition produced in the United States.
Pomorsky, J. (1988, February). *The use of video and west-east flow.* Paper presented at the International Conference on Information Flows Between Eastern and Western Europe, Arnoldsheim, West Germany.
Rislakki, J. (1986, May 19). Finnish tv is Reagan's mouthpiece. *Helsingin Sanomat,* p. 25. Translated by United States Government Joint Publication Research Service-WER-86-065, July 7, 1988, pp. 14.
Schmemann, S. (1983, October 22). Video's forbidden offerings alarm Moscow. *New York Times,* pp. 1, 13.
Shanor, D. R. (1985). *Behind the lines: The private war against Soviet censorship.* New York: St. Martin's Press.

Silinysh, V. (1985, June 30). The video web. *Sovetskaya Latviya*. Document JPRS-UPS-85-066 translated by the Joint Publication Research Service. Washington, DC: U.S. Government.

Smale, A. (1983, April 10). Soviets battle black market in western movie cassettes. *Philadelphia Inquirer*, p. 15.

Special Report: The video revolution in Eastern Europe. (1988, January 20). *Soviet East European Report*, 5(2). New York: Radio Free Europe/Radio Liberty.

Speech by Comrade V. M. Chebrikov, chairman of the USSR committee for state security, (1986, March 1). *Pravda*, pp 5, 6. Reported in British Broadcasting Corporation Monitoring Service, *Daily Report/Soviet Union*, March 3, 1986, pp. 11-13.

Taubman, P. (1985, December, 14-15). Soviet Union trying to control video revolution. *International Herald Tribune*, p. 2.

The wizard of IKI. (1987, October 5). *Time*, p. 69.

Tinsley, E. (1986, April 21). Soviets open video shops. *Electronic Media*, p. G12.

Valko, E. (1985). Video here, video there. *Jelkep*, Mass Communication Research Centre. Budapest, Hungary.

Video equipment. (1985, July 17). *Radio free Europe research,* Hungary. New York: Radio Free Europe/Radio Liberty. (pp. 15-17).

Videos, pirates and the underground. (1986, March). *Index on Censorship*, pp. 18-22.

Walker, C. (1986, April 15). Gorbachov gives go-ahead to the video craze. *The Times* (London), p. 8.

Yasmann, V. (1986, September 22). The collectivization of videos? *Radio Liberty Research Bulletin 355/86*, pp. 1-6. New York: Radio Free Europe/Radio Liberty.

Yasmann, V. (1987, June 29). Satellite television in the USSR: Towards a "new dimension." *Radio Liberty Research Bulletin, RL Supplement 5/87*. New York: Radio Free Europe/Radio Liberty.

About the Contributors

DOUGLAS A. BOYD (Ph.D., University of Minnesota, Minneapolis) is Professor and Dean, College of Communications, University of Kentucky, Lexington. A specialist in international communication research, he has done research in the Arab world, Britain, France, Germany, and the Soviet Union. Professor Boyd is the author of *Broadcasting in the Arab World*: *Radio and Television Broadcasting in the Middle East* (1982), and coauthor of *Videocassette Recorders in the Third World* (1989).

AKIBA A. COHEN is Associate Professor in the Department of Communication and Director of the Smart Family Foundation Communications Institute at the Hebrew University of Jerusalem. Professor Cohen, who received his Ph.D. from Michigan State University, is coauthor of *Almost Midnight*: *Reforming the Late Night News* (1980) and author of *The Television News Interview* (1987).

LAURA COHEN was a master's candidate in Communication at the Hebrew University of Jerusalem, working with Professor Akiba Cohen on a study of videocassette recorder use.

JULIA R. DOBROW is Assistant Professor of Communication at the College of Communication, Boston University. Professor Dobrow earned her A.B. in anthropology and sociology at Smith College and her Ph.D. from the Annenberg School of Communications at the University of Pennsylvania. Her research focuses on the social effects of new communications technologies, and on media use by ethnic and racial groups.

EDWARD FORREST received his Ph.D. from the University of Wisconsin, Madison. Professor Forrest is Associate Professor of Communication and Chair of the Department of Communication at Florida State University in Tallahassee. Dr. Forrest is also coordinator of the Technographics Research Project and director of the New Communication Technologies Laboratory.

BRADLEY S. GREENBERG is Professor of Telecommunication and Communication at Michigan State University, and Chair of the Department of Telecommunication. Professor Greenberg's recent research has focused on the orientation of adolescents and adults to new media in several cultures, including the United States, China, Japan, Great Britain, and Spain. His most recent books are *Patterns of Teletext Use in the United Kingdom* and *Cableviewing*.

BARRIE GUNTER (Ph.D., North East London Polytechnic) is Head of Research at the Independent Broadcasting Authority, London. Dr. Gunter's primary research interests include the impact of television violence, memory and the comprehension of broadcast news, and the influence of television on social beliefs and attitudes. He has written nine books and has published more than 100 papers on various mass communication and psychological topics.

BRUCE C. KLOPFENSTEIN earned his Ph.D. at Ohio State University. Currently an assistant professor in the Radio-Television-Film Department of the School of Mass Communication at Bowling Green State University, Professor Klopfenstein's research interests include the historical and future diffusion of new media technologies, and the impact of the new media on existing media and their audiences.

MARK R. LEVY (Ph.D., Columbia University) is Professor and Associate Dean in the College of Journalism and a Research Associate of the Center for Research in Public Communications at the University of Maryland. He is coauthor of *The Main Source: Learning from Television News* (1986) and *Home Video and the Changing Nature of the Television Audience* (1988) and coeditor of volumes 5 and 6 of the *Mass Communication Review Yearbook*.

CAROLYN LIN is an assistant professor in the Department of Radio-Television at Southern Illinois University, Carbondale. A native of Taiwan, Dr. Lin received her Ph.D. from Michigan State University. Professor Lin's research areas include uses and effects of new media, telecommunications policy, and international communications. Her work has appeared in *Journal of Broadcasting and Electronic Media*, *Journalism Quarterly*, and *Telecommunications Policy*.

THOMAS R. LINDLOF (Ph.D., University of Texas at Austin) is Associate Professor of Telecommunications at the University of Kentucky. Professor Lindlof's research interests include the accommodation of media in the family, interpretive processes of mediated communication, and qualitative research methods. He is editor of *Natural Audiences: Qualitative Research of Media Uses and Effects* (Ablex, 1987).

PAUL B. LINDSTROM is Vice-President/Product Manager for Nielsen Homevideo Index. Mr. Lindstrom is a graduate of New York University. He has spent the past ten years designing and conducting proprietary research on the new media, and he is part of the team which developed the methodology to electronically tract prerecorded cassette playback within the Nielsen PeopleMeter sample.

CHRISTINE OGAN (Ph.D., University of North Carolina at Chapel Hill) is Associate Professor of Journalism at Indiana University, where she teaches courses that examine the national and international implications of communication technologies. Professor Ogan has written extensively on the role of new media technologies in developing countries. She is coauthor of *Newspaper Leadership*, which in part examines technology in the newspaper industry.

KEITH ROE is Associate Professor and Research Fellow at the Unit of Mass Communication, University of Gothenburg, Sweden. Professor Roe's research interests center on the relationship between adolescents' school experience and media use, particularly VCRs, music videos, and popular music. Dr. Roe has been conducting VCR studies since 1981, when he was a member of the Media Panel research team at the University of Lund.

ALAN M. RUBIN (Ph.D., University of Illinois at Champaign-Urbana) is Professor of Journalism and Mass Communication and of Speech Communication at Kent State University. Professor Rubin's research interests include media uses and effects, and the interface of personal and mediated communication.

REBECCA B. RUBIN is Professor of Speech Communication at Kent State University. Professor Rubin received her Ph.D. from the University of Illinois at Champaign-Urbana. Her research interests include interpersonal communication, communication competence, and the personal-mediated communication interface.

BARRY S. SAPOLSKY (Ph.D., Indiana University) is Associate Professor of Communication at Florida State University in Tallahassee. Professor Sapolsky served as Director of the Communication Research Center at Florida State University from 1980 to 1988. His research interests include new information technologies, effects of pornography, and sports spectatorship.

MILTON J. SHATZER is an Assistant Professor in the Department of Telecommunications at the University of Kentucky. He earned his Ph.D. in Communication from Michigan State University in 1987. His teaching and research interests include intercultural and international communication, social effects of telecommunications, the use of telecommunications in development, and research methodologies.

DOV SHINAR is Associate Professor of Communications at Concordia University, Montreal. A former faculty member of the Communications Institute and Director of the Division for Communications in Education of the Hebrew University, Professor Shinar is author of *Communications and Aging: Bringing the Mountain to Mohammed* and *Palestinian Voices: Communication and Nation Building in the West Bank*.

MALLORY WOBER (Ph.D., London) has done applied psychological research in Africa, in the study of individual adjustment in girls' boarding schools and in hostels for retarded adolescents. For the past fifteen years, Dr. Wober has pioneered in the use of subjective measures of assessment of broadcast use. He is coauthor of *Television and Social Control* and author of *The Use and Abuse of Television: A Social Psychological Analysis of the Changing Screen*.

HD 9696.V532 V37 1989

DATE DUE